SHARKS

First edition
Published by Reader's Digest Services Pty Ltd (Inc. in NSW)
26-32 Waterloo Street, Surry Hills, NSW 2010

Edited and designed by Capricorn Press Pty Ltd (Inc. in NSW)
3/9 Oaks Avenue, Dee Why, NSW 2099

National Library of Australia
cataloguing-in-publication data
Sharks, silent hunters of the deep
    Includes index
    ISBN 0 86438 014 3
    1. Sharks. I. Reader's Digest Services.
597'.31

SHARKS
Silent hunters of the deep

Reader's Digest
Sydney London New York Montreal Cape Town

Contributors

Alice Alston

Ian Close, BA(Hons)

Perry W. Gilbert, PhD
is Director Emeritus of the Mote
Marine Laboratory, Sarasota,
Florida, and Professor Emeritus of
Neurobiology and Behaviour at
Cornell University, Ithaca, New
York. He has studied the biology
and behaviour of sharks in many
parts of the world for the past 45
years, and is currently working
with the sensory and reproductive
systems of sharks.

Peter Goadby
is a world famous Australian big
game fisherman with a special
interest in sharks. He is a regular
contributor to many international
publications and has one of
Australia's best private collections
of rare books on fish and fishing.

Robert Pullan

Olaf Ruhen

**Julian G. Pepperell,
BSC(Hons), PhD**
has worked in the field of marine
recreational fisheries for over 10
years. He is a senior biologist with
the Fisheries Research Institute,
NSW Department of Agriculture,
and specialises in the study of large
pelagic predatory fishes. He is
responsible for the Australian
Gamefish Tagging Programme.

Ron and Valerie Taylor
are perhaps the best known
husband and wife diving team.
Their work with sharks,
particularly the large, dangerous
species, has given them more
practical experience in this field
than anyone else in the world.

**Susan Turner, PhD,
AMA, FGS, FLS**
is an Honory Research Fellow of the
Queensland Museum. She has
studied the ancient jawless fishes –
the thelodonts – for 20 years, and
the similarity of their scales to
those of modern sharks led to her
study of cartilaginous fishes. Since
1980 she has been researching fossil
sharks from Australia and China.

C. Scott Johnson, PhD
was trained as a physicist, and has spent the past 23 years studying hearing and echolocation in dolphins, shark behaviour, the electrical senses of sharks and various countermeasures to the dangerous species.

Noel R. Kemp, DDA, DipEd, MSc, MAIG
has been Curator of Geology at the Tasmanian Museum since 1973. Among his particular interests are the teeth of sharks from the last 70 million years, the taxonomy of these fossils, and their relationships to modern species.

Nick Otway, BSc(Hons)
is conducting marine biological research for his doctorate at the University of Sydney. His research interests lie in the field of experimental marine ecology, and he has participated in several shark tagging studies while working with the CSIRO.

Design
Lawrence Hanley

Art
Alistair Barnard

Head of a grey nurse shark.

Contents

8   Introduction by Ron and Valerie Taylor

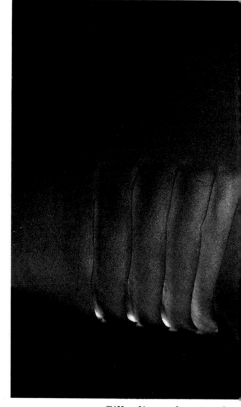

Gills slits and pectoral
fin of a grey nurse shark.

Head of a grey nurse shark
Eugomphodus taurus.

World of the shark
12   Realm of the shark
20   The mysterious lives and habits of sharks
32   The shark's remarkable senses
44   Sharks in the wild
50   Great white death
54   The largest fish on earth

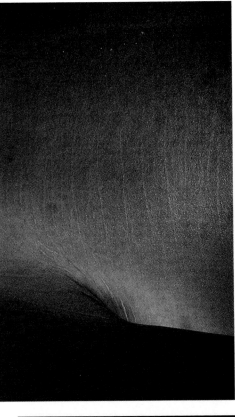

Men and sharks

60 'A marveilous strange fishe'
72 The risk of attack
82 Testing a new theory
84 Tales of terror and bravery
88 The persistent shark
96 A bolt from the blue
104 The final parade
106 Where sharks are gods
112 The shark papers
114 The shark arm mystery
120 Death in New Jersey
126 Not all sharks are killers
132 The shark and the Lord Mayor
136 Keeping sharks at bay
142 The suit of steel
144 Shark for sale

Denticles glisten on the skin of a grey nurse shark.

Facts about sharks and shark attack

154 Sharks of the world
180 Facts, fallacies and records
186 Global patterns of shark attack
198 General index
202 Species index
208 Acknowledgements

INTRODUCTION

Few creatures on earth are as much feared, and as little understood, as sharks. Lions eat people; they are given special reserves and the title 'King of the Beasts'. Sharks eat people; their name has become a byword for ferocity, and people demand their extermination.

For centuries sharks have been merely objects of terror. A real understanding of their lives and habits only came with the invention and development of free-diving equipment after World War II. We were among the first to film sharks underwater in the 1950s, and have taken an active part ever since in bringing the extraordinary wonders of the ocean world to a wider audience through film, television and still photography.

We hope that this magnificent book will help readers to see sharks as we do – to appreciate the important role that they play in the complex web of underwater life. Certainly large sharks can be terrifying creatures. They are formidable and efficient predators and demand respect. Certainly people are occasionally injured, or even killed by them, and few deaths can be as terrible. We ourselves have been frightened on occasions during the 30 years we have worked closely with large, so called 'maneating', sharks. We have been jostled and bumped by them as they snatched food from the water around us, yet we have only ever received minor, accidental injuries. The truth is that most sharks ignore people – they are neither friends nor enemies.

Many eminent, internationally renowned scientists, researchers, writers and photographers have contributed to this important book. All are experts in their own fields, and over the years we have had the pleasure of working with many of them in their attempts to unravel the mysteries of shark biology and behaviour. Their contributions have brought together all the important facts about sharks, in what is certainly the most informative, entertaining and beautifully illustrated book ever published on this subject.

Many people will be drawn to this book by the terrifying stories it contains. We hope that they will come away with a greater appreciation of a sadly misunderstood group of complex and fascinating creatures.

Ron &
Valerie Taylor

A sinister shape, lurking in the shadows – two divers confront the ultimate underwater nightmare. Despite its menacing appearance, this grey nurse shark is quite harmless.

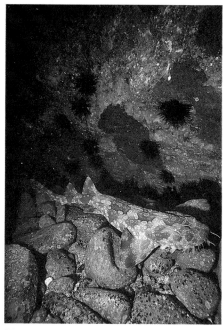

Sharks are an extraordinarily diverse group of creatures – only a minority conform to the popular, sleek, 'maneater' image. Illustrated here are a wobbegong (above), a Port Jackson shark (right) and a hammerhead (below), all from Australian waters.

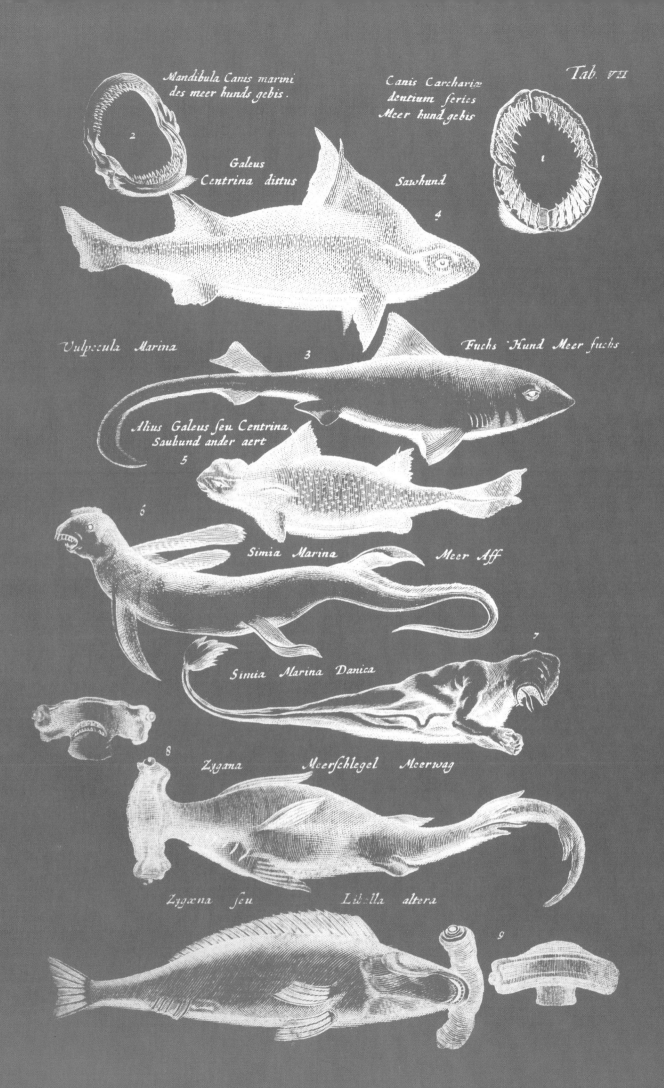

Tab. VII

Mandibula Canis marini
des meer hunds gebis.

Canis Carchariæ
dentium feries
Meer hund gebis

Galeus
Centrina dictus

Sawhund

Vulpecula Marina

Fuchs Hund Meer fuchs

Alius Galeus feu Centrina
Sauhund ander aert

Simia Marina

Meer Aff

Simia Marina Danica

Zygæna

Meerschlegel   Meerwag

Zygæna feu

Libella altera

Part One

1

World of the Shark

Sharks are difficult
to study. Many details
of their lives remain a
mystery to this day. The
extraordinary behaviour of this
ancient group of animals and
their strange abilities are
only now beginning to be
understood by scientists.

Illustrations of sharks
from Conrad Gesner's
Fisch-Buch, first
published in 1670.

REALM OF THE SHARK

For most people the watery world inhabited by sharks is almost as alien as the surface of another planet. Even the most experienced of divers are usually restricted to nearshore waters. The great ocean depths – little known and little explored – still harbour many mysteries.

The earth, viewed from far above the Pacific Ocean, deserves the name given to it by astronauts: the blue planet. This one ocean alone covers 181.3 million sq. km (70 million sq. miles), and is 17 700 km (11 000 miles) across at its widest.

The motions of the atmosphere, traced out by clouds, and the size of the oceans dominate the view of earth from space. So vast are the oceans, in fact, that they take up almost 71 per cent of the entire surface of the globe. The combined area of the oceans and seas is 361 million sq. km (139 million sq. miles).

The oceans of the earth have an average depth of 3730 m (12 230 ft), and reach their deepest point in the Marianas Trench in the north-western Pacific Ocean at 11 038 m (36 204 ft) below sea level. The ocean basins hold at least 1185 million cu. km (285 cu. miles) of water. Scientists believe that this vast quantity of water arose from the earth's interior as it cooled.

The water in the oceans and seas is a solution of many elements and chemical compounds which give it its distinctive salty taste. After some early fluctuations, the chemical composition of seawater stabilised, and it has remained more or less the same since early Palaeozoic times, about 600 million years ago.

Approximately 2730 million tonnes of dissolved solids are poured into the oceans each year by rivers. This material is almost enough to account for the saltiness of seawater, but analysis shows that there are more of some elements in the oceans than there are in river water, so there must be another source of dissolved minerals.

This additional source was not found until about the middle of this century when the mechanisms responsible for continental drift were investigated. Meandering through the ocean basins are the mid-ocean ridges where fresh rock flows up from deep in the earth, causing the sea floor to spread. Accompanying this rock are virgin waters rich in salts and trace-elements that are less abundant in water that runs off the land.

The total quantity of dissolved solids in seawater – its salinity – is about 35 parts per thousand, and oceanic water generally ranges from 34 to 37 parts per thousand. Variations from place to place are due to factors such as rainfall, evaporation, biological activity and radioactive decay.

Fresh supplies of salts are now being added to the oceans at roughly the same rate they are being removed by various physical, chemical and biological processes.

Measures of salinity do not include the dissolved gases in ocean water, many of which are important to marine life. All of the gases found in the atmosphere are also present in seawater, and the major ones in air – nitrogen, oxygen, argon and carbon dioxide – are also the most abundant gases in the oceans. The gases come from a variety of sources. Nitrogen, for example, enters the ocean dissolved in rainwater, whereas oxygen is produced by marine plants during photosynthesis, and also enters through the aerating action of waves. Factors such as temperature, salinity, depth, the state of the sea surface and biological activity all affect the distribution and abundance of gases.

The temperature of the oceans varies enormously according to season, latitude and depth. The warmest waters are at the surface near the equator where they average about 28°C (82.4°F). The coldest waters are found in the depths of the Arctic and Southern Oceans where the average is -2°C

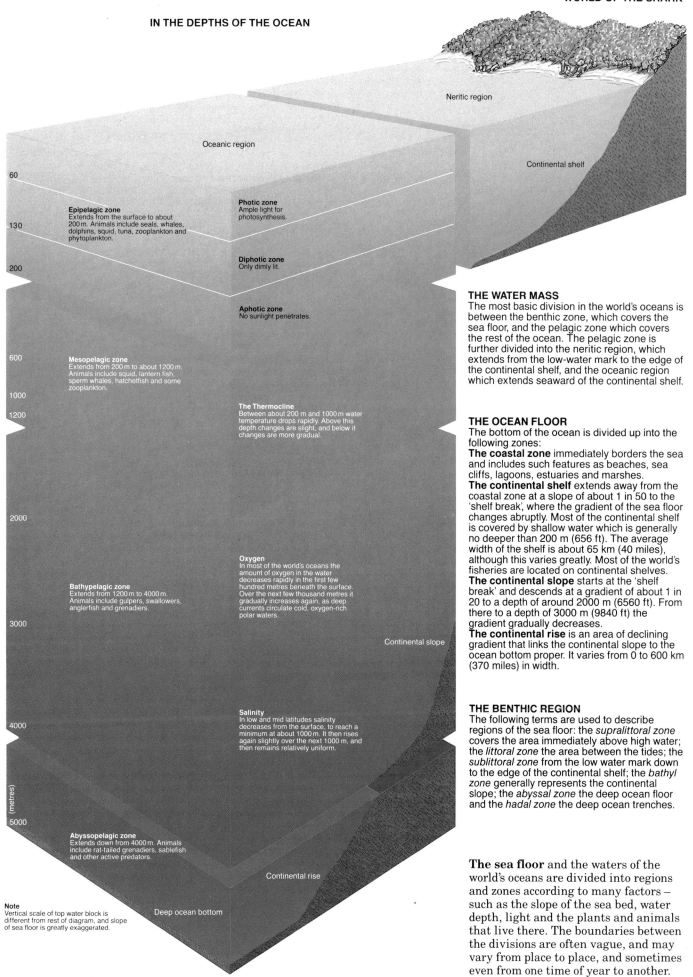

IN THE DEPTHS OF THE OCEAN

Neritic region

Oceanic region

Continental shelf

60

Epipelagic zone
Extends from the surface to about
200 m. Animals include seals, whales,
dolphins, squid, tuna, zooplankton and
phytoplankton.

Photic zone
Ample light for
photosynthesis.

130

200

Diphotic zone
Only dimly lit.

Aphotic zone
No sunlight penetrates.

600

Mesopelagic zone
Extends from 200 m to about 1200 m.
Animals include squid, lantern fish,
sperm whales, hatchetfish and some
zooplankton.

1000

1200

The Thermocline
Between about 200 m and 1000 m water
temperature drops rapidly. Above this
depth changes are slight, and below it
changes are more gradual.

2000

Oxygen
In most of the world's oceans the
amount of oxygen in the water
decreases rapidly in the first few
hundred metres beneath the surface.
Over the next few thousand metres it
gradually increases again, as deep
currents circulate cold, oxygen-rich
polar waters.

Bathypelagic zone
Extends from 1200 m to 4000 m.
Animals include gulpers, swallowers,
anglerfish and grenadiers.

3000

Continental slope

Salinity
In low and mid latitudes salinity
decreases from the surface, to reach a
minimum at about 1000 m. It then rises
again slightly over the next 1000 m, and
then remains relatively uniform.

4000

5000

(metres)

Abyssopelagic zone
Extends down from 4000 m. Animals
include rat-tailed grenadiers, sablefish
and other active predators.

Continental rise

Note
Vertical scale of top water block is
different from rest of diagram, and slope
of sea floor is greatly exaggerated.

Deep ocean bottom

THE WATER MASS
The most basic division in the world's oceans is
between the benthic zone, which covers the
sea floor, and the pelagic zone which covers
the rest of the ocean. The pelagic zone is
further divided into the neritic region, which
extends from the low-water mark to the edge of
the continental shelf, and the oceanic region
which extends seaward of the continental shelf.

THE OCEAN FLOOR
The bottom of the ocean is divided up into the
following zones:
The coastal zone immediately borders the sea
and includes such features as beaches, sea
cliffs, lagoons, estuaries and marshes.
The continental shelf extends away from the
coastal zone at a slope of about 1 in 50 to the
'shelf break', where the gradient of the sea floor
changes abruptly. Most of the continental shelf
is covered by shallow water which is generally
no deeper than 200 m (656 ft). The average
width of the shelf is about 65 km (40 miles),
although this varies greatly. Most of the world's
fisheries are located on continental shelves.
The continental slope starts at the 'shelf
break' and descends at a gradient of about 1 in
20 to a depth of around 2000 m (6560 ft). From
there to a depth of 3000 m (9840 ft) the
gradient gradually decreases.
The continental rise is an area of declining
gradient that links the continental slope to the
ocean bottom proper. It varies from 0 to 600 km
(370 miles) in width.

THE BENTHIC REGION
The following terms are used to describe
regions of the sea floor: the *supralittoral zone*
covers the area immediately above high water;
the *littoral zone* the area between the tides; the
sublittoral zone from the low water mark down
to the edge of the continental shelf; the *bathyl
zone* generally represents the continental
slope; the *abyssal zone* the deep ocean floor
and the *hadal zone* the deep ocean trenches.

The sea floor and the waters of the
world's oceans are divided into regions
and zones according to many factors –
such as the slope of the sea bed, water
depth, light and the plants and animals
that live there. The boundaries between
the divisions are often vague, and may
vary from place to place, and sometimes
even from one time of year to another.

Invertebrates are one of two major divisions of the animal kingdom. The name means 'without a backbone', but apart from this distinction most have little else in common. The group ranges from plankton to jellyfish and from worms to prawns.

Vertebrates – animals with backbones – include all the most highly evolved creatures on earth. The most numerous vertebrates in the oceans are the bony fishes, but the group also includes whales, dolphins, seals, turtles and of course sharks and rays.

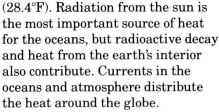

Zooplankton

Galapagos shark

Sea star

Hermit crab

Jellyfish

Loggerhead turtle

Coral cod

Fur seal

(28.4°F). Radiation from the sun is the most important source of heat for the oceans, but radioactive decay and heat from the earth's interior also contribute. Currents in the oceans and atmosphere distribute the heat around the globe.

Temperature and salinity are critical to life in the oceans because they can affect the rate at which chemical reactions take place, including those involving the enzymes essential to all living cells. As a consequence, many animals can only live in water where temperature and salinity fluctuate slightly. Others, however, can tolerate conditions which vary a good deal.

Many of the salts in the oceans, especially the calcium salts, are used to build animal skeletons. The shells of molluscs, the spines and tests (cases) of sea urchins and the teeth and bones of fish are all made up mostly of calcium carbonate absorbed from water. Nitrates and phosphates, also present in sea-water, are essential to plant growth and are easily extracted and used. The concentrations of these chemical compounds can directly affect the number of phytoplankton – tiny floating plants – in the water.

A final vital component of seawater, and an important source of food for many ocean dwellers, are the millions of tiny particles of organic matter that come from dead and decaying plants and animals. A constant rain of these particles sinks gradually to the ocean floor, often to be returned to the surface by currents.

The energy that drives the ocean's circulation comes almost exclusively from the sun. Surface currents transport enormous volumes of warm water to the polar regions, while deep currents move water from one hemisphere to the other. On a smaller scale, storms and strong winds may cause local upwellings of cold water.

Ocean currents also play an important role in distributing gases throughout the ocean. Descending currents at the poles, for example, take oxygen-rich surface water to the ocean depths where it is used by the animals living on or near the sea bed.

Light is extremely important to life in the oceans because it is needed by plants for photosynthesis – the process by which sugar is created from carbon dioxide and water, using light for energy.

Sunlight can only penetrate to a maximum depth of 600 m (1970 ft) in clear seawater, although the actual amount of light at any depth can vary from place to place for many reasons, including reflection from the water surface, refraction, and the murkiness of the water. This means that marine plants can only live in the upper part of the oceans where there is sufficient light for photosynthesis. The distribution of plants in turn governs the areas where plant-eating animals can live. Light also affects the behaviour of animals in the ocean – many rise to the surface to feed at night.

The oceans can be divided into two basic regions, the benthic and the pelagic – the former covering the ocean floor and the latter the rest of the water mass. Organisms that live in the pelagic zone are known either as nekton – the active, free-swimming animals – or plankton – small plants and animals that float or drift passively near the water surface. There are two types of plankton – the phytoplankton, made up of millions of microscopic plants, and the zooplankton, tiny animals which include the eggs and larvae of many of the ocean's larger inhabitants.

Phytoplankton are the primary producers of the ocean. They convert energy from the sun into living plant tissue, and thus form the basis of most food chains. As much as 80 to 90 per cent of all the photosynthetic activity that takes place on earth is carried out in the upper 100 m (328 ft) of the oceans.

Zooplankton are usually much larger than phytoplankton, although many are less than 10 mm (0.4 in) long. Most of them can swim, although they are usually only able to travel up and down through the water. Zooplankton generally feed on phytoplankton, and they are therefore the second step in the food chain, coverting plant into animal tissue. Zooplankton are, in turn, preyed on by larger animals such as fish, squid and some sharks.

Most free-swimming animals in the oceans are vertebrates – they have backbones – such as sharks,

OCEAN FOOD CHAINS

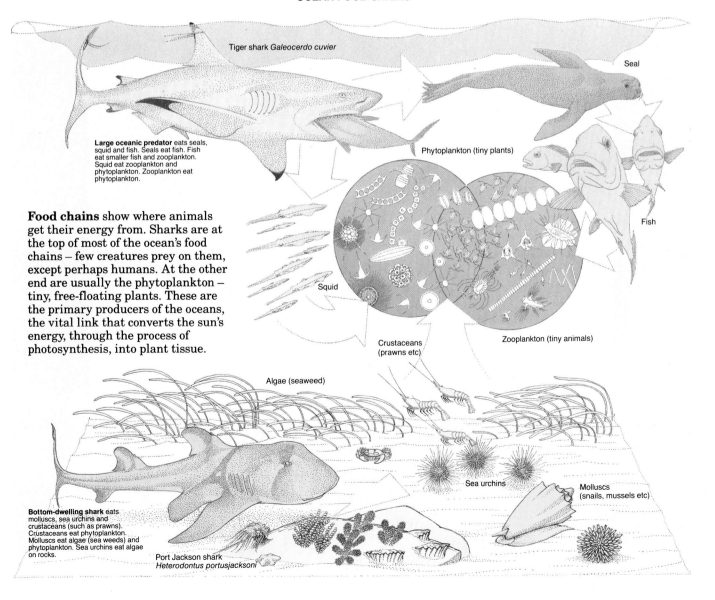

Tiger shark *Galeocerdo cuvier*

Seal

Large oceanic predator eats seals, squid and fish. Seals eat fish. Fish eat smaller fish and zooplankton. Squid eat zooplankton and phytoplankton. Zooplankton eat phytoplankton.

Phytoplankton (tiny plants)

Fish

Food chains show where animals get their energy from. Sharks are at the top of most of the ocean's food chains – few creatures prey on them, except perhaps humans. At the other end are usually the phytoplankton – tiny, free-floating plants. These are the primary producers of the oceans, the vital link that converts the sun's energy, through the process of photosynthesis, into plant tissue.

Squid

Zooplankton (tiny animals)

Crustaceans (prawns etc)

Algae (seaweed)

Sea urchins

Molluscs (snails, mussels etc)

Bottom-dwelling shark eats molluscs, sea urchins and crustaceans (such as prawns). Crustaceans eat phytoplankton. Molluscs eat algae (sea weeds) and phytoplankton. Sea urchins eat algae on rocks.

Port Jackson shark *Heterodontus portusjacksoni*

Bony fishes have evolved along quite a different path to that taken by the sharks, and they are the most successful of the ocean's vertebrates. There are at least 20 000 (known) species alive today, and they may be found in salt and fresh water, in almost every part of the globe.

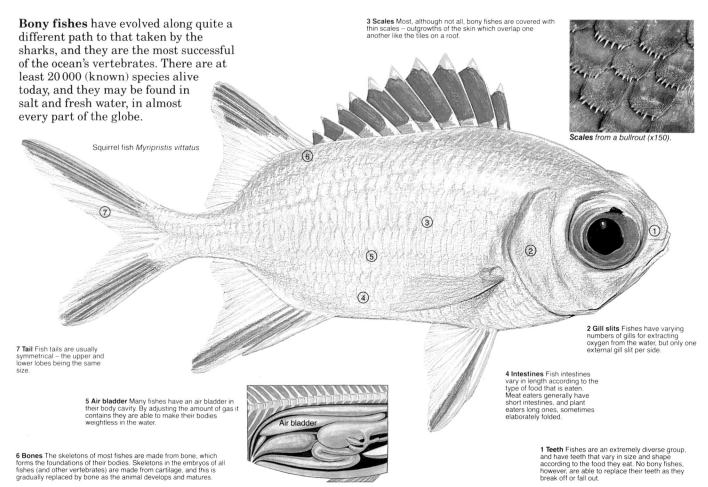

Squirrel fish *Myripristis vittatus*

3 Scales Most, although not all, bony fishes are covered with thin scales – outgrowths of the skin which overlap one another like the tiles on a roof.

Scales from a bullrout (x150).

7 Tail Fish tails are usually symmetrical – the upper and lower lobes being the same size.

5 Air bladder Many fishes have an air bladder in their body cavity. By adjusting the amount of gas it contains they are able to make their bodies weightless in the water.

Air bladder

6 Bones The skeletons of most fishes are made from bone, which forms the foundations of their bodies. Skeletons in the embryos of all fishes (and other vertebrates) are made from cartilage, and this is gradually replaced by bone as the animal develops and matures.

2 Gill slits Fishes have varying numbers of gills for extracting oxygen from the water, but only one external gill slit per side.

4 Intestines Fish intestines vary in length according to the type of food that is eaten. Meat eaters generally have short intestines, and plant eaters long ones, sometimes elaborately folded.

1 Teeth Fishes are an extremely diverse group, and have teeth that vary in size and shape according to the food they eat. No bony fishes, however, are able to replace their teeth as they break off or fall out.

tuna, herring and whales. These animals are generally carnivorous, preying on zooplankton or other vertebrates. In contrast to these free-swimming animals, there are those that live on the sea floor which are either permanently attached to rocks, or are relatively slow moving. They include crabs, lobsters, abalone, oysters, starfish, sea urchins, corals and sponges. The plants that live on the sea floor, such as green, brown and red algae (seaweeds), are restricted to areas that are reached by enough sunlight for photosynthesis.

Many of the animals that live in the oceans share a phase of their life cycles, when their eggs or larvae, or both, become part of the zooplankton. They are then food for other zooplankton and many free-swimming species, and are also completely at the mercy of ocean

currents which can transport them over great distances.

The free-swimming animals of the oceans are dominated by the fishes, sharks and rays. The structure of a fish body is designed for ease of movement, and this is especially true of the sharks. This ability to move easily, without having to rely on water currents to carry them about, has enabled fishes to exploit most parts of the world's oceans, and this is reflected in an extraordinary variety of sizes and shapes.

There are many species of bony fishes, sharks and rays, and an understanding of the relationships between them is based on detailed studies of their anatomy, development, and on the sometimes scanty fossil record. What is known of the earliest fishes has been gleaned from fragments of bone taken from

rocks of the early Ordovician age, which are approximately 480 million years old. Some researchers think that the origins of fishes may go back further in time because the two major groups – the Agnatha or jawless fishes, and the Gnathostomata or jawed fishes – were so well established at the beginning of the Silurian period, about 435 million years ago.

The most primitive of living marine vertebrates are the cyclostomes – the lampreys and hagfishes – which are greatly modified descendants of the Agnatha. They have a skeleton made of cartilage – a tough, elastic tissue – instead of bone, and do not have jaws. These slimy, eel-like animals are adapted primarily for scavenging and for life as parasites.

Modern jawed fishes fall into two groups – the cartilaginous

ETWEEN SHARKS AND FISHES

Sharks are, in many ways, more highly evolved than many of the bony fishes, although they have been less successful in terms of numbers and distribution. There are 344 (known) species and they are found almost exclusively in salt water, although some occasionally venture into rivers and lakes.

Blacktip reef shark
Carcharhinus melanopterus

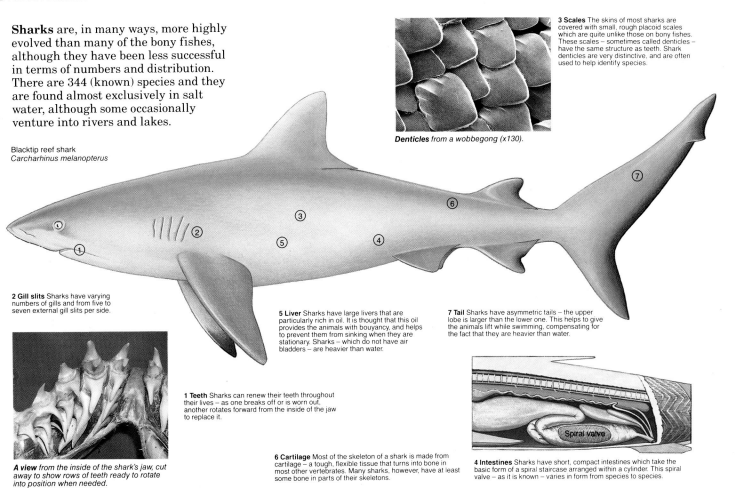

3 Scales The skins of most sharks are covered with small, rough placoid scales which are quite unlike those on bony fishes. These scales – sometimes called denticles – have the same structure as teeth. Shark denticles are very distinctive, and are often used to help identify species.

Denticles from a wobbegong (x130).

2 Gill slits Sharks have varying numbers of gills and from five to seven external gill slits per side.

5 Liver Sharks have large livers that are particularly rich in oil. It is thought that this oil provides the animals with bouyancy, and helps to prevent them from sinking when they are stationary. Sharks – which do not have air bladders – are heavier than water.

7 Tail Sharks have asymmetric tails – the upper lobe is larger than the lower one. This helps to give the animals lift while swimming, compensating for the fact that they are heavier than water.

A view from the inside of the shark's jaw, cut away to show rows of teeth ready to rotate into position when needed.

1 Teeth Sharks can renew their teeth throughout their lives – as one breaks off or is worn out, another rotates forward from the inside of the jaw to replace it.

Spiral valve

6 Cartilage Most of the skeleton of a shark is made from cartilage – a tough, flexible tissue that turns into bone in most other vertebrates. Many sharks, however, have at least some bone in parts of their skeletons.

4 Intestines Sharks have short, compact intestines which take the basic form of a spiral staircase arranged within a cylinder. This spiral valve – as it is known – varies in form from species to species.

AN ANCIENT PAST

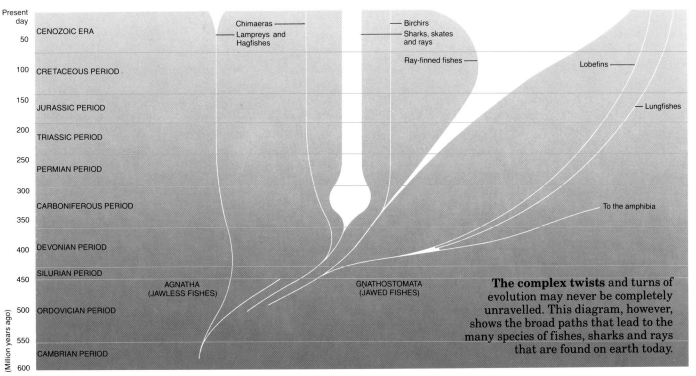

Present day

Chimaeras
Lampreys and Hagfishes
Birchirs
Sharks, skates and rays
Ray-finned fishes
Lobefins
Lungfishes
To the amphibia

CENOZOIC ERA	50
CRETACEOUS PERIOD	100
JURASSIC PERIOD	150
TRIASSIC PERIOD	200
PERMIAN PERIOD	250
CARBONIFEROUS PERIOD	300 / 350
DEVONIAN PERIOD	400
SILURIAN PERIOD	450
ORDOVICIAN PERIOD	500
CAMBRIAN PERIOD	550 / 600

AGNATHA (JAWLESS FISHES)

GNATHOSTOMATA (JAWED FISHES)

(Million years ago)

The complex twists and turns of evolution may never be completely unravelled. This diagram, however, shows the broad paths that lead to the many species of fishes, sharks and rays that are found on earth today.

fishes (Chondrichthyes) such as sharks, rays and rat-fishes (chimaeras), and the bony fishes (Osteichthyes). There are more than 20 000 species of bony fishes which are divided into four main groups – the lungfishes, the lobefins, the ray-finned fishes (which make up the majority) and the birchirs and reedfishes. Bony fishes are found in both the oceans and in fresh water, a habitat which has never been successfully colonised by the sharks and rays. Most fishes are carni-

vorous, but a few feed on algae, such as seaweeds.

The cartilaginous fishes are divided into two sub-classes – the Elasmobranchii or sharks, skates and rays, and the Holocephali or rat-fishes (chimaeras), rabbit-fishes and elephant-fish. Fishes in this last sub-class live close to the ocean floor and have retained many primitive features. They have no stomach; their teeth are in large plates firmly attached to their jaws; they have small apertures for their

mouths; and their upper jaws are fused to their skulls. They feed on molluscs, other invertebrates and sometimes small fishes.

The sharks, skates and rays are not quite the masters of the oceans that the bony fishes are, but they are well adapted to their environment and survive in large numbers. Almost all are carnivores or scavengers, although the species that live close to the sea floor feed mostly on invertebrates. Three distinctive features identify all

Living sharks are divided into eight orders – groups that are used by scientists to collect together animals that have certain common characteristics. One member from each order is illustrated here. It is not, however, possible to illustrate one shark that is 'typical' of each order, because their external appearance can vary greatly. The carpet sharks (*Orectolobiformes*), for example, include the whale shark – the world's largest fish – and the wobbegongs, patterned sharks that live on the sea floor.

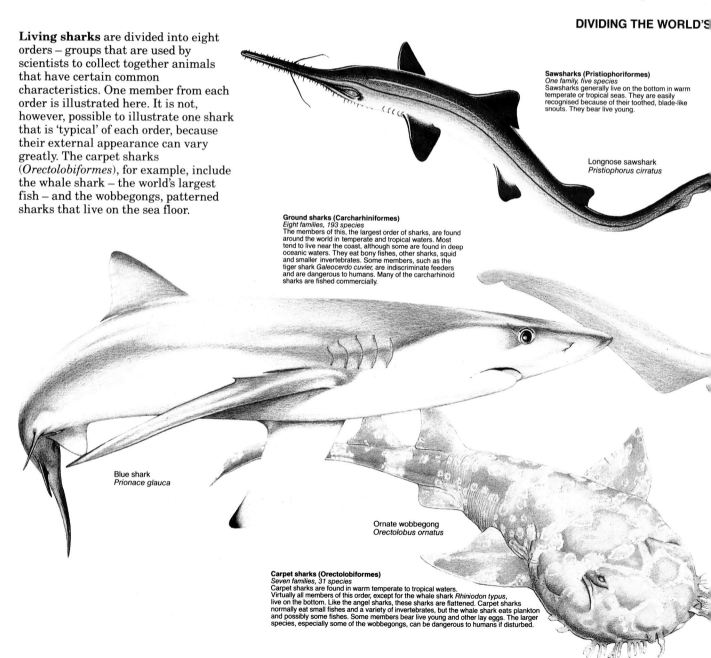

Sawsharks (Pristiophoriformes)
One family, five species
Sawsharks generally live on the bottom in warm temperate or tropical seas. They are easily recognised because of their toothed, blade-like snouts. They bear live young.

Longnose sawshark
Pristiophorus cirratus

Ground sharks (Carcharhiniformes)
Eight families, 193 species
The members of this, the largest order of sharks, are found around the world in temperate and tropical waters. Most tend to live near the coast, although some are found in deep oceanic waters. They eat bony fishes, other sharks, squid and smaller invertebrates. Some members, such as the tiger shark *Galeocerdo cuvier*, are indiscriminate feeders and are dangerous to humans. Many of the carcharhinoid sharks are fished commercially.

Blue shark
Prionace glauca

Ornate wobbegong
Orectolobus ornatus

Carpet sharks (Orectolobiformes)
Seven families, 31 species
Carpet sharks are found in warm temperate to tropical waters. Virtually all members of this order, except for the whale shark *Rhiniodon typus*, live on the bottom. Like the angel sharks, these sharks are flattened. Carpet sharks normally eat small fishes and a variety of invertebrates, but the whale shark eats plankton and possibly some fishes. Some members bear live young and other lay eggs. The larger species, especially some of the wobbegongs, can be dangerous to humans if disturbed.

sharks, skates and rays – their skeletons are made of cartilage and not bone; their skin is covered with denticles and not scales; and they have either five or seven gill slits per side, and not one per side like all bony fishes.

Sharks are superbly adapted to their environment. Most possess a keen sense of smell, a large brain, good eyesight and a highly specialised mouth and teeth. Their bodies are usually heavier than water and they do not have an air-filled swim bladder for buoyancy like most bony fishes. All sharks have an asymmetric tail fin, with the upper lobe being larger than the lower one, although this is less noticeable in fast-swimming species. This feature, together with flattened pectoral fins, and an oil-filled liver compensates for the lack of a swim bladder.

There are 344 known species of sharks (to date) living in all parts of the oceans, from shallow to deep water, and from the tropics to the polar regions. A few even venture into fresh water and have been found in rivers and lakes. Contrary to popular belief, most sharks are harmless to humans.

Sharks are classified into eight orders (see illustration below). Unfortunately very little is known about the general biology and behaviour of most of the world's sharks. Research is concentrated almost exclusively on the few species considered to have commercial potential.

SHARKS INTO EIGHT ORDERS

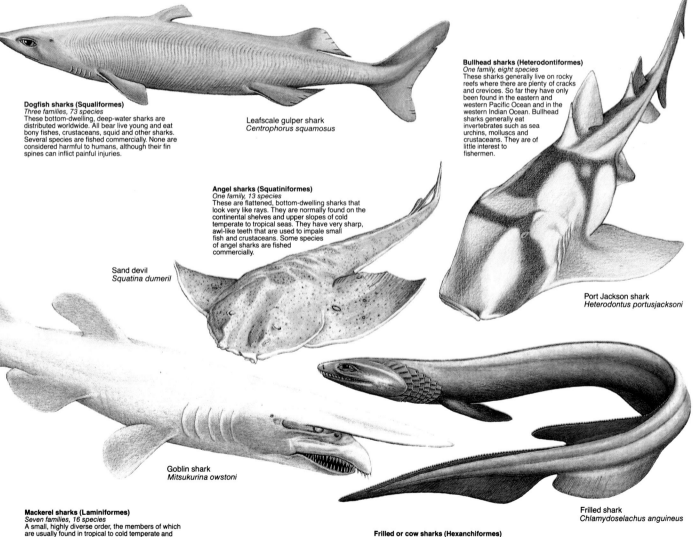

Dogfish sharks (Squaliformes)
Three families, 73 species
These bottom-dwelling, deep-water sharks are distributed worldwide. All bear live young and eat bony fishes, crustaceans, squid and other sharks. Several species are fished commercially. None are considered harmful to humans, although their fin spines can inflict painful injuries.

Leafscale gulper shark
Centrophorus squamosus

Bullhead sharks (Heterodontiformes)
One family, eight species
These sharks generally live on rocky reefs where there are plenty of cracks and crevices. So far they have only been found in the eastern and western Pacific Ocean and in the western Indian Ocean. Bullhead sharks generally eat invertebrates such as sea urchins, molluscs and crustaceans. They are of little interest to fishermen.

Angel sharks (Squatiniformes)
One family, 13 species
These are flattened, bottom-dwelling sharks that look very like rays. They are normally found on the continental shelves and upper slopes of cold temperate to tropical seas. They have very sharp, awl-like teeth that are used to impale small fish and crustaceans. Some species of angel sharks are fished commercially.

Sand devil
Squatina dumeril

Port Jackson shark
Heterodontus portusjacksoni

Goblin shark
Mitsukurina owstoni

Mackerel sharks (Laminiformes)
Seven families, 16 species
A small, highly diverse order, the members of which are usually found in tropical to cold temperate and even Arctic waters. Some are oceanic while others are found near the coast. Most grow quite large and eat bony fishes, other sharks, squid and marine mammals. The basking sharks and megamouth, however, are plankton eaters. This order includes some partially warm-blooded sharks, such as the mako and great white, both of which are known to attack humans.

Frilled or cow sharks (Hexanchiformes)
Two families, five species
These are primarily deep-water, bottom-dwelling sharks with a worldwide distribution. They are easily recognised because of their six or seven gill slits (all other sharks have five pairs, except for the sawshark Pliotrema, which has six). They bear live young and generally eat bony fishes, crustaceans and other sharks.

Frilled shark
Chlamydoselachus anguineus

THE MYSTERIOUS LIVES AND HABITS OF SHARKS

Sharks are hard to study. Many species, including the ones most dangerous to humans, cover great distances in their daily routines. In order to see them at all it is usually necessary to put bait in the water, and then only feeding behaviour is observed. Our present knowledge of sharks is about equivalent to our understanding of other animals at the beginning of the twentieth century.

In spite of the difficulties involved, a great deal has been learned about sharks and their cousins, the rays, skates, and chimaeras (fishes with large heads and long, tapering tails), in the past several years. The general view of sharks is changing. Once thought of as stupid – a swimming nose attached to an eating machine, a fish that should have disappeared with the dinosaurs – they are now viewed as highly evolved and efficient predators with brains, in proportion to body size, larger than many birds and mammals. Sharks can learn faster than bony fishes, at a rate comparable to white rats and pigeons. At various laboratories young nurse sharks *Ginglymostoma cirratum* have been trained to come when signalled, to touch paddles and other objects and, on command, to retrieve a ring that is thrown into their holding tank.

There are great differences in size and shape among the 344 species of sharks known to exist. The largest fish in the world is the whale shark *Rhiniodon typus* which reaches 12 m (40 ft) long, and the smallest shark is the dwarf shark *Squaliolus laticaudus* which matures at a length of 150 mm (6 in). Sharks inhabit all the oceans of the world from the Arctic to the Antarctic. They frequent, if not inhabit, many rivers that flow into the oceans, and at least one lake, Lake Nicaragua. Actually only one species, the bull shark *Carcharhinus leucas*, is known to frequent fresh water routinely. At one time it was thought that there were three species of fresh water sharks: the Zambezi River shark, the Ganges River shark, and the Lake Nicaragua shark, because they were found frequently in the lake and the two rivers. It has now been determined that they are all the same species, the bull shark. The bull shark is one of the most dangerous to humans, having been responsible for many attacks in all parts of the world.

The migratory patterns of several shark species have been studied, and temperature is a major environmental factor affecting their movements. Some sharks are known to be warm blooded: the three species of thresher sharks *Alopias vulpinus*, *A. superciliosus* and *A. pelagicus*, the great white shark, two species of mako sharks *Isurus oxyrinchus* and *I. paucus*, the salmon shark *Lamna ditropis*, and the porbeagle shark *L. nasus*. But most sharks are cold blooded. The benthic species, living at the bottom in deep water, are in an environment that is probably never more than a few degrees above 0°C (32°F). Northern hemisphere pelagic species, seen near the surface in deep water, tend to migrate north as the water temperature rises in the spring and summer, and south as the temperature drops in autumn and winter. The reverse is true in the southern hemisphere. These movements are also affected by warm and cold currents. The great blue shark *Prionace glauca* likes

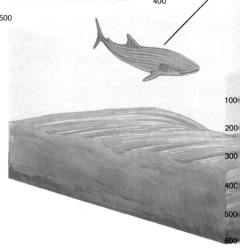

400
500
100
200
300
400
500
600

Only two specimens of the remarkable megamouth shark *Megachasma pelagios* have ever been captured. Both were males, 4.5 m (14.8 ft) long. One was caught near the Hawaiian Islands and the other off the coast of California. Megamouth, like the whale and basking sharks, is a filter feeder, and yet anatomically it is related more closely to the great white shark *Carcharodon carcharias*.

SHARKS AND OCEAN ZONES

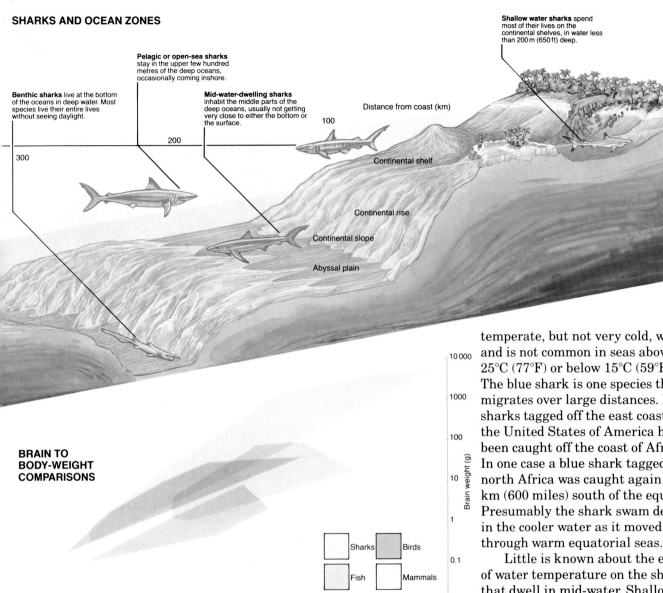

Shallow water sharks spend most of their lives on the continental shelves, in water less than 200 m (650 ft) deep.

Pelagic or open-sea sharks stay in the upper few hundred metres of the deep oceans, occasionally coming inshore.

Benthic sharks live at the bottom of the oceans in deep water. Most species live their entire lives without seeing daylight.

Mid-water-dwelling sharks inhabit the middle parts of the deep oceans, usually not getting very close to either the bottom or the surface.

Distance from coast (km)

300

200

100

Continental shelf

Continental rise

Continental slope

Abyssal plain

BRAIN TO BODY-WEIGHT COMPARISONS

Brain weight (g)

10 000
1000
100
10
1
0.1
0.01

| Sharks | Birds |
| Fish | Mammals |

Body weight (kg)

0.001　0.01　0.1　1　10　100　1000　10 000　100 000

Claims that sharks are stupid are not borne out by comparisons of brain to body-weight ratios among a range of animals. Some sharks have larger brains, in proportion to body weight, than most bony fishes, many birds and some mammals.

Proof of the ability of sharks to learn has come from experiments such as those conducted in the United States of America with these nurse sharks, which were conditioned to press a target in order to receive food.

temperate, but not very cold, water and is not common in seas above 25°C (77°F) or below 15°C (59°F). The blue shark is one species that migrates over large distances. Blue sharks tagged off the east coast of the United States of America have been caught off the coast of Africa. In one case a blue shark tagged off north Africa was caught again 965 km (600 miles) south of the equator. Presumably the shark swam deeper in the cooler water as it moved through warm equatorial seas.

Little is known about the effects of water temperature on the sharks that dwell in mid-water. Shallow water sharks, as is the case with pelagic sharks, move in response to temperature changes. In the tropics where temperature changes are not great, the sharks may stay in a relatively limited area for most of their lives. Some shallow water sharks, such as nurse sharks, do not appear to migrate with changing temperatures – they simply become less active as the water gets cooler. Food supply as well as temperature affects movements.

Because all sharks receive their oxygen supply from the water via their gills they tend to avoid very warm water – above 32°C (90°F) - which has a lower oxygen content than cooler water. High water purity is also important, particularly for the long term health of captive sharks. Sharks, as is the case with bony fishes, do not survive well in polluted water.

Sharks, specially large sharks,

ANATOMY OF A SHARK

The bodies of sharks vary enormously – there is no 'typical' shark. However, the following descriptions of this shortfin mako Isurus oxyrinchus *also apply to most other species.*

1 Eye *Most shark eyes have immovable lids, but a few species (not the mako) have an eyelid – the nictitating membrane – that can completely cover the eye to protect it.*

2 Nostril *Shark nostrils are very sensitive and in tests have responded to extremely low concentrations of some chemicals.*

3 Ampullae of Lorenzini *Sensors within these small pores on the heads of sharks detect minute electric fields, and are used to find prey buried under sand.*

4 Gill slits *Sharks have from five to seven gill slits per side.*

5 Spiracle *These apertures – one per side – behind the eyes of most sharks are vestigal gill slits which play a role in aerating blood.*

6 Teeth *Shark teeth vary greatly in shape according to the way in which the animal feeds. The thin, sharp teeth of makos are adapted for seizing the fish that they eat.*

7 Skull *The shark's skull is a compact block of cartilage containing the brain as well as the auditory and nasal capsules.*

8 Gill arches *These support the gills.*

9 Auditory capsule *A shark's ears are hidden within its skull and are connected to the exterior of the animal by narrow canals.*

10 Skeleton *The skeleton of a shark is made up almost entirely of cartilage – a tough flexible tissue.*

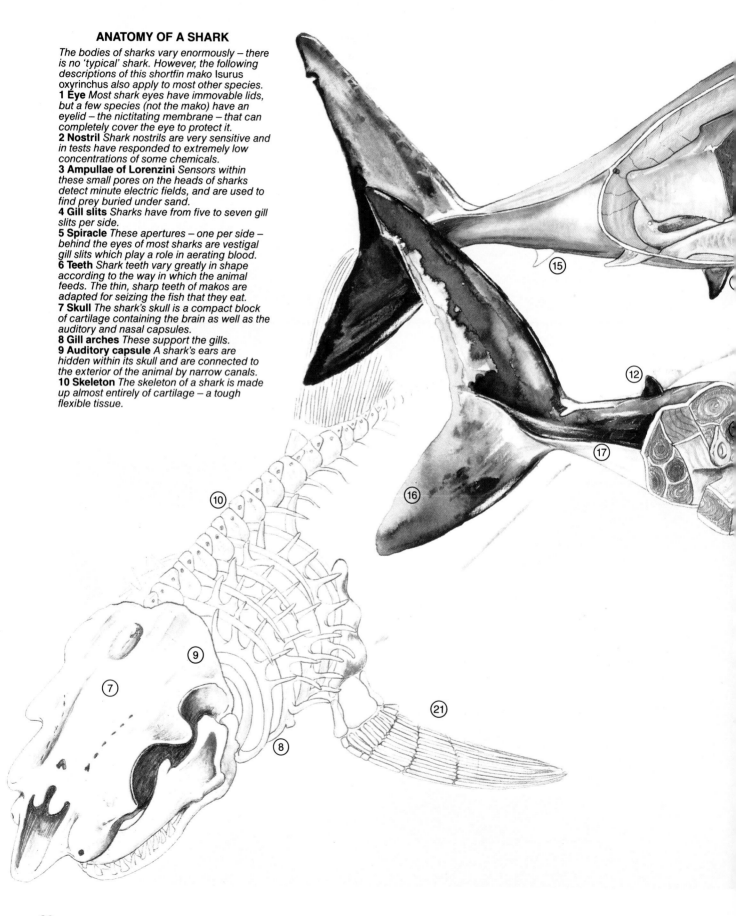

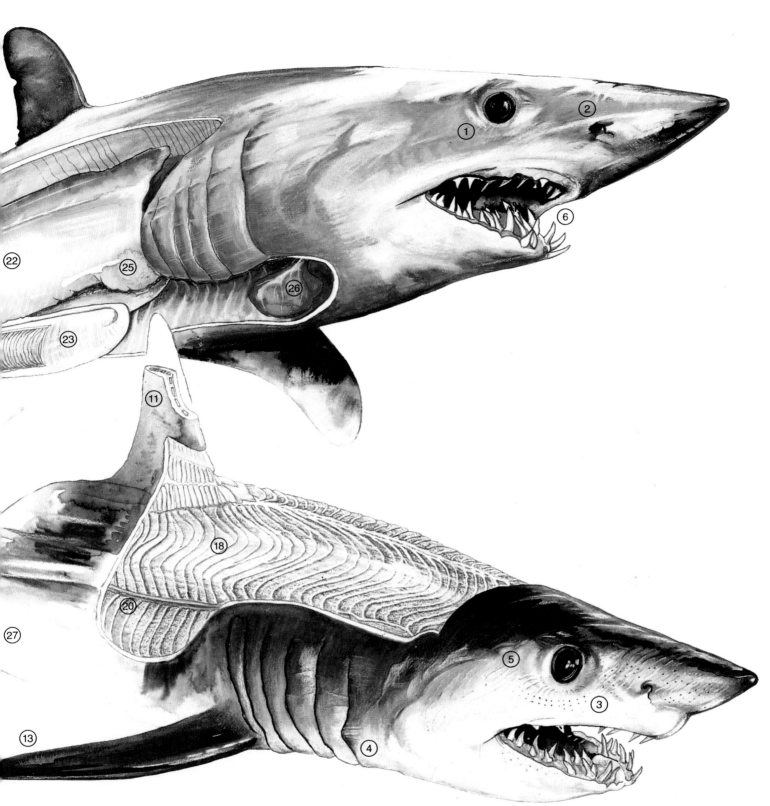

11 Dorsal fin
12 Second dorsal fin
13 Pectoral fin
14 Pelvic fin
15 Anal fin
16 Caudal fin *Most sharks have assymetric tails – the upper lobe being larger than the lower. This is less noticable in fast swimming sharks such as the mako.*
17 Caudal keel *The bodies of many fast swimming sharks are slightly flattened at the back which may help them to move quickly. However, this feature is also present in some slow moving species, such as the plankton-eating whale shark.*
18 Muscle structure

19 Dark muscle *A few fast swimming sharks (including the mako) have a body temperature that is higher than the surrounding water. The reason for this is that warm muscle is more efficient than cold. The dark, warm muscle is used by the animal for swimming continuously at cruising speed. Heat generated in the muscle is used to warm cold blood coming from the gills.*
20 Lateral cutaneous artery *This artery is the major source of blood only in warm-bodied sharks such as the mako and porbeagle.*
21 Fin structure *It is the horny rods used to support the fin internally that are dried and extracted for use in soup.*

22 Liver *Shark livers have two lobes which almost completely enclose the digestive tract. In a large tiger shark the oil-rich liver can weigh up to 80 kg (176 lb).*
23 Spiral valve *The intestine in a shark is short and compact. Internally it is like a spiral staircase arranged within a cylinder.*
24 Stomach *Shark stomachs are U-shaped.*
25 Pancreas
26 Heart
27 Skin *Shark skin is rough because it is covered with tiny denticles. These small, hard scales have the same internal structure as the animal's teeth, and are unique to sharks and rays. Denticles vary in shape from species to species.*

Nurse sharks *Ginglymostoma cirratum* are one of several species that are known to rest on the sea floor. At one time it was thought that sharks had to keep moving in order to extract enough oxygen from the water.

are surprisingly fragile. While it is sometimes very difficult to kill a shark quickly, it is easy to injure one during capture and transport. Injured sharks frequently die in a few days or weeks. Many injuries result from the fact that sharks have no bony skeleton to support their vital organs. On the other hand, disease is almost unknown in sharks. Most sharks have parasites, both internal and external, but none is life-threatening. The greatest threat to their lives comes from humans and other sharks.

Sharks live fairly long lives. The spiny dogfish *Squalus acanthias* is estimated to live more than 30 years. This small and harmless shark grows very slowly, only reaching a maximum length of 1 to 1.5 m (39 to 59 in). Females are sexually mature at about 12 years, and males at 9 years.

It is not known what factors determine the maximum size of sharks, but, as in humans and other mammals, growth hormones may perform this function. In the case of one of the larger species, the tiger shark *Galeocerdo cuvier* which grows to over 5 m (16.4 ft), it has been suggested that their maximum size may be limited by the size of their livers. These contain large quantities of oil which provide the shark with buoyancy (see box story far right).

Another shark, the sandtiger *Eugomphodus taurus*, which often swims very slowly, virtually hanging still in the water, may have developed an alternative way of acquiring buoyancy. Sandtigers sometimes come to the surface and appear to gulp air into their stomachs, and this may help to prevent them from sinking to the bottom when they are not moving.

At one time it was thought that many species of sharks never slept or stopped swimming for very long because they had weak gill muscles, and had to keep moving to get

oxygen from the water. The only ones that appeared to sleep were the nurse sharks, bullhead sharks, cat-sharks, carpet sharks and a few others. As time progresses, and more observations are made of both captive and wild sharks, the number of species that are known to stop swimming and lie on the bottom for extended periods is increasing. Bull sharks and grey reef sharks *Carcharhinus amblyrhynchos* have both been observed to do this. So far none of the pelagic sharks have been seen resting on the sea floor.

Sharks swim by moving their tails from side to side in an undulatory motion, like many other fish. All sharks have asymmetric tails, the upper lobe of the tail being larger than the lower lobe. Many bottom dwelling sharks have no lower lobe at all. A larger upper tail lobe is needed to provide downward thrust to balance the lift

Most sharks are hosts to a variety of parasites – both internal and external — like these sea lice living in the mouth of a great white shark. Despite the fact that some sharks carry enormous numbers of parasites, especially internal parasites such as tapeworms, they seem to be little affected by them. Sharks tolerate remoras, or sharksuckers, because they feed on external parasites.

generated by pectoral fins at the front of the shark so that they can swim level. The great white shark, and all the warm blooded sharks, have the most nearly symmetrical tails, except for the thresher shark. Thresher sharks are asymmetric in the extreme. They do have a small lower lobe, but the upper lobe makes up about 40 per cent of their overall length. The long tail is used to strike the water surface and stun bait fish so that they are easier to catch and eat.

Sharks generally cruise around at a relatively slow one to two metres per second (0.7 mph), but some are capable of short bursts of relatively high speeds. One mako shark has been estimated to have exceeded 32 km/h (20 mph) for a few seconds. Sharks tire quickly under the stress of fast swimming. The most powerful swimmers are the warm blooded sharks. This is because warm muscles are stronger

Shark tails differ in size and shape according to the species. In every case, however, the upper lobe is larger than the lower lobe so that the shark can create downward thrust to counteract the lift generated by the pectoral fins.

Upper lobe

Lower lobe

Tiger sharks are slower swimmers than porbeagles and whites, and also have markedly asymmetrical tails.

Nurse sharks spend much of their time lying on the sea floor. Their tails have almost no lower lobe, and consequently they swim in an eel-like manner.

Porbeagle sharks are fast swimmers and have more-nearly symmetrical tails than other species. They are capable of rapid acceleration.

Thresher sharks have the most extraordinary tails, with the upper lobe making up 40 per cent of the length of the shark's body.

Great white sharks also have almost symmetrical tails, which seem to be a major source of thrust when swimming.

Cookiecutter sharks have large tails, probably so they can accelerate rapidly to catch their prey – whales, dolphins and large fish.

than cold ones. A 3°C (6°F) increase in temperature doubles muscle strength, and muscle temperatures of up to 8°C (16°F) above the surrounding water have been measured. These warm blooded species also have smoother skins than other sharks, thus reducing water drag and increasing swimming efficiency. The skins of most sharks are designed for protection, not drag reduction. Sharks are not adapted for sustained swimming at high speeds, it uses too much energy. Rather they are designed for long periods of slow cruising. The drag on a brown shark *Carcharhinus plumbeus* has been measured at eight to ten times that of a similar sized Pacific whiteside dolphin *Lagenorhynchus obliquidens*.

Compared to other fish, sharks are not particularly powerful. A 1.5-m (4.9-ft) bull shark was measured to develop a maximum

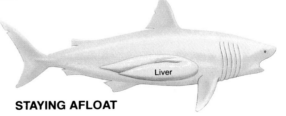

Basking sharks need a large liver to provide them with enough buoyancy to remain just beneath the water surface for long periods of time.

Liver

STAYING AFLOAT

The size of a shark's liver may be the key to its overall length. Unlike many other fishes, sharks do not have air-filled swim bladders to give them buoyancy. They are therefore heavier than water by a few per cent, and must produce lift in order to swim off the bottom of the sea.

Like aeroplanes in air, sharks 'fly' underwater using their pectoral fins like wings. If the fuselage of an aircraft is doubled in size, then the wing area must be more than doubled if the 'plane is to fly. This is because a doubling in size cubes the volume of the fuselage, but only squares the area of the wings.

In tiger sharks the fin area of a two-metre-long (6.6-ft) specimen is just half that of a 4-m (13.1-ft) shark. In order to compensate for the loss of lift caused by having a bulkier body and relatively smaller fins, the 4-m (13.1-ft) shark increases its liver size by considerably more than twice that of the

2-m (6.6-ft) shark. The liver contains oil, and it is this that gives the larger shark the increased buoyancy that it needs.

Sharks obviously cannot keep giving over a larger and larger proportion of their body cavity to house the liver, otherwise there would be no room for other organs. This theory therefore predicts that there should be a maximum size for any species of shark. Calculations so far have only been made for tiger sharks, and these suggest that the maximum length for that species should be around 5 to 6 m (16 to 20 ft), a figure which agrees with records of large specimens captured so far.

The livers of large sharks can reach a great weight – sometimes up to 91 kg (200 lbs). In the days when sharks were fished for the oil in their livers it was reckoned that a well-fed, 4-m (13-ft) tiger shark might contain a liver that would yield as much as 81.8 litres (18 gallons) of rich oil.

The swirling bodies of a group of grey reef sharks *Carcharhinus amblyrhychos* in a feeding frenzy present an awesome spectacle. In these circumstances sharks will attack anything around them, even one another.

Sharks have teeth adapted to deal with their food. The Port Jackson shark, for example, has small, pointed teeth at the front for seizing prey, and large molars at the back for crushing the shells of the molluscs it eats.

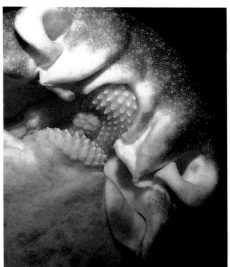

swimming power of about 0.2 kW (0.25 horsepower), while a 1.2-m (3.9-ft) barracuda *Sphyraena barracuda* produced a maximum of more than 0.6 kW (0.75 horsepower). Barracuda are adapted for remaining almost still and then accelerating quickly to high speed to catch prey.

The mating of all sharks is much like that of many mammals: the male introduces sperm directly into the female's reproductive tract. Males have two organs, called claspers, for doing this, but probably only one is used at a time. The claspers are extensions of the male's pelvic fins. Pre-mating behaviour and actual copulation have been observed in only a few species. In some instances mating appears to be a rather violent activity. Adult females have been seen with lacerations on their pectoral fins and on their backs and sides between the dorsal fin and tail, where the males bite them before mating. In some species the males have smaller teeth than females, and female blue sharks have skins that are twice as thick as males in the critical areas. The best guess as to how males identify receptive females is through their extremely acute sense of smell.

Shark young are brought into

the world in three different ways, with some variations. As with most mammals, many sharks bear live young which are nurtured in one of the female's two oviducts through umbilical cords. No care is given to the young after they are born. Some sharks lay eggs, and as with the live bearers, once the eggs are laid they receive no future care. The whale shark lays pillow-shaped eggs almost 500 mm (20 in) long. One of the most interesting egg layers is the Port Jackson shark *Heterodontus portusjacksoni* from Australia. The eggs of this species are shaped like large truncated wood screws 150 mm (5.9 in) long, with spiral flanges on the outside. Some scientists think that the female carries the eggs around in her mouth after they are laid, and deposits them in reef crevices where, because of the shape, they become securely lodged.

In the third method of bearing young, the females produce eggs and carry them internally until the young hatch. There is no attachment to the mother. The young are nourished by their large egg yolks. An unusual case is that of the sandtiger shark which produces hundreds of pea-sized eggs throughout a long gestation period of several months. One of the first

eggs in each oviduct develops into a baby shark before the others, and each baby sustains itself until birth by cannibalizing its undeveloped brothers and sisters as they are produced by the mother. Sandtigers only produce one large baby shark – 40 per cent of the length of the mother – from each of the two oviducts. While probing the uterus of a pregnant sandtiger a scientist was bitten by an unborn pup. While not the smallest, this was definitely the youngest shark to have attacked a human! The incident has been classified as a provoked attack.

The gestation period in sharks varies from about nine months to as long as twenty-two months for the spiny dogfish. The number of young produced varies from species to species, but ranges from one to close to a hundred in the tiger shark.

In spite of the fact that inanimate objects are frequently found in shark's stomachs, they are all entirely carnivorous. Food ranges from the smallest plankton to the largest whales. Sharks prefer live fresh food but do sometimes eat decaying flesh. There are, however, no vultures in the shark family that specialize in eating only dead flesh. A favourite food is apparently shark pups. When sharks are first caught and brought into captivity it is

12 FD SQN RE

3 TROOP

These teeth all come from large, fast-swimming, flesh-eating sharks, and they all have the same basic characteristics – they are pointed and extremely sharp. Teeth from the upper and lower jaws of the same shark are often different shapes because of the way in which such sharks feed. Prey is first impaled on the teeth of the lower jaw, which are generally pointed, and those in the upper jaw are used to 'saw' away a section of flesh. The serrated edges on the teeth of some species help with this process.

Although sharks' teeth are sharp, they are also brittle, and are constantly breaking off. Fortunately for the sharks, new ones are being created all the time, and these rotate quickly into position from the inside of the jaw as they are needed.

THE TEETH OF A PREDATOR

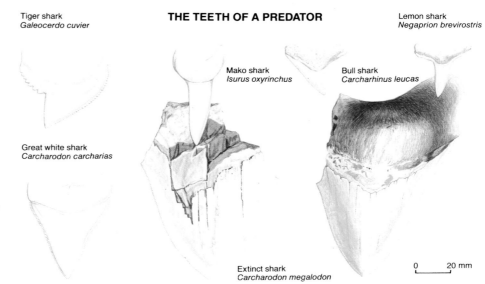

Tiger shark
Galeocerdo cuvier

Lemon shark
Negaprion brevirostris

Mako shark
Isurus oxyrinchus

Bull shark
Carcharhinus leucas

Great white shark
Carcharodon carcharias

Extinct shark
Carcharodon megalodon

0   20 mm

often difficult to get them to take food. This can sometimes be remedied by giving them live baby sharks or fresh shark liver. It is thought that bull sharks have their pups in fresh water to protect their young from being eaten by other sharks. Some bottom-dwelling sharks, such as the Port Jackson shark, feed extensively on crustaceans and shellfish. These sharks have molars at the backs of their mouths for crushing shells, and sharp teeth at the front for seizing small prey.

Among the sharks that are dangerous to humans, the shape of the teeth ranges from long, narrow and pointed, as in the mako shark, to broad and somewhat less pointed in the case of the tiger shark. In every case, however, the teeth have edges that are razor-sharp – quite sharp enough to easily shave the hairs from an arm. In many of the dangerous species the teeth in the upper jaw are broader and more triangular than those in the lower jaw. In all shark jaws there are several sets of teeth – five or more – present at the same time. Behind the set of teeth visible and being used by the shark are the other sets, folded back into the jaw tissue. When a sharks loses a tooth, a replacement rotates out of the jaw

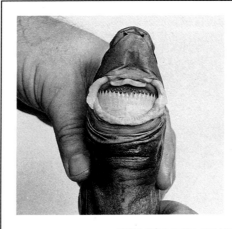

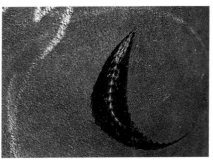

The razor sharp teeth of the little (500-mm; 20-in) cookiecutter shark (left) can cause considerable damage, even to the rubber casing on submarine components (above).

SOLVING THE COOKIECUTTER RIDDLE

When submarines surface after dives there is the ever-present danger of collision with one of the growing number of surface ships plying the oceans of the world. To help avoid this problem United States Navy submarines have a listening device that can be raised far enough above the boat to hear surface ships. This listening device is covered with a 13-mm (0.5-in) rubber coating. About 10 years ago it was found that these rubber coatings were being damaged. Crescent shaped slashes were somehow being cut out of them.

At first it was thought that the damage was due to some kind of animal bite, but since none of the biologists that were asked could identify the animal responsible, they were assumed to be due to mechanical failure. Exhaustive tests could not reproduce the slashes. Finally, quite by accident, a scientist who studied both whales and sharks was shown some of the slashes. 'Of course,' he said, 'these are caused by the cookiecutter shark which feeds by biting chunks out of whales and

large fish. The cookiecutter must mistake submarines for whales!'

Once the source of the damage was identified it was a simple matter to cover the rubber coating with a layer of fibreglass too hard for the little shark to bite through.

Some 30 submarines had been damaged in this way. Fortunately, none of the sharks had ever bitten all the way through the rubber coating. If this had happened, the oil in the listening device would have leaked out, making it useless, and a collision might have resulted. A 500-mm (20-in) long shark could have caused the loss of a submarine.

Despite their small size, cookiecutters are fierce fighters and many fishermen have reported extensive damage to nets in which the sharks have been caught. The injuries to the cookiecutter's living victims are distinctive and take the form of a 50-mm (2-in) wide hole where flesh and skin have been gouged out. Until quite recently the resulting scars on whales and dolphins were thought to be caused by a bacterial infection, or some sort of parasite.

tissue to take its place. In small, growing sharks, each tooth is replaced once every one to two weeks. In adults this rate probably slows to replacement every month or two. Even so, every shark must produce hundreds of sets of teeth in its lifetime.

Not all sharks lose teeth one or two at a time. Some species lose a whole set at once. An example of this is the cookiecutter or cigar shark *Isistius brasiliensis*. This little shark, which grows to 500 mm (20 in) in length, lives in the mid-water volume in deep tropical oceans. It eats by biting chunks the size and shape of half a tennis ball out of whales, tuna, squid and other larger animals. Its upper jaw contains several rows of needle sharp teeth which protrude 1 mm (0.04 in) from the jaw. In the lower jaw are 25 to 30, 3 mm (0.12 in) long teeth that are strongly attached to one another, forming a band. The jaws are surrounded by muscular

lips which form a suction attachment when it attacks its prey (see box p.27). Once attached to the smooth surface of its victim, the needle-like teeth in the upper jaw prevent the shark from slipping off, and it then scoops out a hemispherical chunk of flesh. On occasions the shark swallows its teeth at the same time as its meal. Complete sets of teeth have been found in the stomachs of many captured specimens.

The cookiecutter shark is also one of the several species of sharks that have luminescent organs on the outsides of their bodies. These may be used to attract prey in the deep, dark waters they inhabit. Megamouth sharks *Megachasma pelagios* have luminous organs inside their mouths which may be used to attract the small fish and squid they filter from the water.

Because sharks have such sharp teeth they do not need extremely strong jaw muscles. Even so,

measurements have shown that some species can generate impressive forces when they bite. Up to 587 newtons (132 pounds-force) have been measured between the jaws of a 2-m (6.6-ft) dusky shark *Carcharhinus obscurus*. Because this force is concentrated at the fine cutting edge of the shark's teeth, it is amplified to 3 tonnes per sq. cm (21 tons per sq. in), which is enough to severely damage oceanographic instruments and other important equipment, not to mention fragile human bodies.

Sharks that live high enough in the ocean to be exposed to sunlight feed mostly at night and are especially active on dark, moonless nights. Their eyes are adapted to see well in low light levels. All sharks also have electrical receptors which detect the weak electric field that surrounds all animals, including humans. Sharks have the highest sensitivity to electric fields of any known marine animal, and

Copulation has been observed in very few sharks, particularly in the wild. The best observations are probably those made of nurse sharks which will mate in captivity, but it is not clear if their mating behaviour is 'typical'.

Male sharks have two claspers with which to introduce sperm into the female cloaca, in the same way as a penis is used in mammals. Opinion is divided on whether one or both of the claspers are used during mating, as both possibilities have been reported by researchers. Claspers vary in shape according to the species (those below are of a great white shark), and are equipped with hooks and spurs, presumably to keep them in place.

Pre-mating behaviour in many sharks

is highly complex, perhaps because a high degree of cooperation is required during copulation. Males will often bite females on their fins and bodies, possibly to stimulate a sexual response, and the wounds created are often severe (below). In nurse sharks insertion of a (single) clasper and copulation takes only about a minute or so.

The corkscrew-shaped eggcase of a Port Jackson shark *Heterodontus portusjacksoni*. Some researchers believe that the female carries the egg cases in her mouth to underwater rocks, and then inserts them into a crevice. Their shape makes them difficult to dislodge. The eggcases of other shark species come in many shapes and sizes.

this gives them a great advantage over their prey. They can find crabs and fish buried in the sand, and detect fish and other creatures in complete darkness, when many animals are not active (see p.35).

It is generally thought that if dolphins are sighted, the area must be free of sharks, because sharks and dolphins are thought to be mortal enemies. Such is not the case. Sharks and dolphins have been seen swimming together on many occasions. There are many reported instances of dolphins attacking and killing sharks. Dolphin remains have been found in sharks and many live, healthy dolphins have been captured bearing large scars which were probably caused by shark bites. The conditions under which these violent encounters occur are not known. Experiments have been conducted in which captured sharks and bottlenose dolphins *Tursiops truncatus* have been placed together in the same enclosure. Under these conditions the behaviour of both species indicated awareness of each other's presence, but there was no aggressiveness shown by either. The sharks tended to swim around the perimeter of the enclosure, while the dolphins kept to the centre. Both groups of animals competed for food without incident.

At the Mote Marine Laboratory at Sarasota, Florida, a bottlenose dolphin was trained to chase and attack sharks. These experiments showed that bottlenose dolphins can tell one shark species from another, for the dolphin would only attack the species with which it had been trained. It ignored any others.

Sharks have few natural enemies, apart from other sharks. The only predator large enough to tackle a shark is a killer whale, although no encounters have been reported. A shark was, however, once taken from the stomach of a sperm whale.

Blackie (above) chasing young blacktip sharks through the shallow waters of Canton Island's lagoon. The dog and his owner, Tonga John (left), solved a pressing scientific puzzle.

THE SHARK-CATCHING DOG OF CANTON ISLAND

The study of sharks in the wild can present researchers with many unexpected problems, not the least of which is that of cornering a suitable specimen.

Canton Island is a small atoll in the Phoenix group, located about 965 km (600 miles) north of Samoa, 1290 km (800 miles) west of the Gilbert Islands and about 3220 km (2000 miles) south-west of Hawaii. This group of islands has now been handed over to the Gilbert Islanders, and is part of the tiny nation of Kiribati.

Canton Island was brought to the attention of scientists from Hawaii and California several years ago by the United States Air Force. The Air Force had based a tracking station on the island for monitoring missile and space flight tests. For reasons unknown to this day, an 8-m (26.5-ft) female whale shark had swum into Canton's shallow lagoon, and Air Force officials feared that the great fish was trapped and would die if not rescued. The shark stayed for over a year, despite all efforts to set her free. Residents of the island named her Mimi, and on days off would swim with her and feed her shrimps and fish. However, she apparently became bored with all the attention and one day swam away and was never seen again.

During trips to the island scientists noticed that Canton's reefs were a treasure-trove of marine life, and several new species of fish were discovered. There was also an abundant supply of juvenile, 450 to 600 mm (1.5 to 2 ft) long, blacktip sharks *Carcharhinus melanopterus* — a species that is commonly found around many Pacific ocean reefs and islands.

These small sharks swam in very shallow water near the shore where their larger cousins could not make a meal of them. It appeared to be a simple task to net the sharks and transport them live to various aquariums and oceanariums for display and study.

As it turned out, the sharks had other ideas, and after three exhausting days spent chasing sharks around in the equatorial heat only six had been caught. By now the catching crew had dwindled to one scientist – all the rest found more pressing work to do. In desperation the remaining scientist enlisted a local civilian Air Force employee from Tonga, aptly named Tonga John.

The island had been deserted between the end of World War II and recent Air Force occupation, except for several dogs which had been left behind when the last inhabitants had departed. A descendant of these dogs, named Blackie, had adopted Tonga John.

The scientist, Tonga John and Blackie went off to collect sharks. When they arrived at the appropriate spot the scientist explained to Tonga John the procedure used to catch sharks. John was not very enthusiastic about the method, but agreed to give it a try. They picked up the collecting nets and turned to find a squirming shark at their feet, with a wet dog, one foot holding the shark down, wagging his tail furiously. The scientist asked where the shark came from. 'Oh', John replied, 'Blackie, he catchum, dats his favourite ting, catchin sharks.' For the next hour Tonga John sat in the shade sipping beer while Blackie, assisted by the scientist, caught the last eight sharks needed. Everyone was happy: the scientist got his sharks plus a great story for cocktail hour, Tonga John got a day off and a few free beers, and Blackie, well, Blackie got to do two of his favourite things – catch sharks and eat all the ice cubes that he wanted.

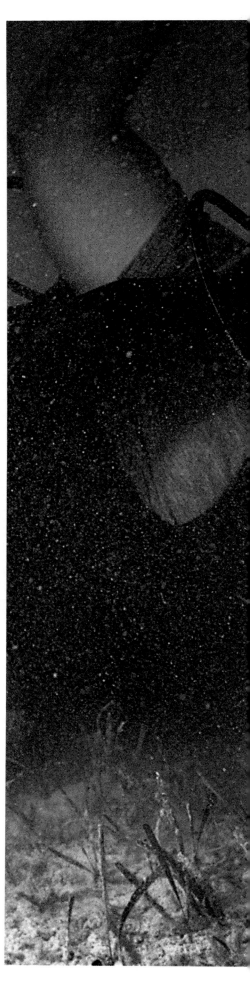

A unique sequence of photographs (reading down from the top) show the actual birth of a lemon shark pup *Negaprion brevirostris*. This event, which had never been seen in a shark in the wild before, took place in Bimini Lagoon in the Bahamas. The mother – which was about 2.6 m (8.5 ft) long – had been captured for study and was anchored to the scientist's boat by a hook in her mouth and a rope around her tail. As each pup was born it would lie briefly on the ocean floor before swimming away, straining against the umbilical cord until it broke, leaving a small depression in the baby shark's belly. The remoras swimming around the mother would dart in rapidly after each pup was born to devour the afterbirth. Altogether 10 pups were born, each about 610 mm (2 ft) long, although the last was born dead. Once free of their mother these baby sharks must survive on their own.

THE SHARK'S REMARKABLE SENSES

Sharks have a range of sensory systems on their heads and bodies with which to explore the environment, pursue prey, and avoid predators. In addition to touch, taste and a keen sense of smell, sharks possess excellent vision at close range and unique vibration and electromagnetic senses. Armed with such equipment, it is not surprising that they have survived so successfully for more than 200 million years.

The shark has frequently been referred to as a 'swimming nose' and there is considerable support for this claim. The nostrils, located on the underside of the snout just ahead of the mouth, open into spacious sacs lined with many folds of tissue which contain the cells responsible for detecting smells. When a shark takes water into its mouth to constantly aerate its gills, suction causes some water to flow in and out of each nasal sac. In addition, the forward motion of the shark brings water through the funnel-shaped nostrils into the sacs. A fleshy flap extending across the opening of each sac separates water flowing in from water flowing out. The odour-detecting system of a shark is therefore constantly bathed by a current of water whether the shark is swimming or resting. In one of the oddest of all sharks, the hammerhead, the nostrils, as well

as the eyes, are located far apart at the ends of the 'hammer'. This fact, coupled with the shark's habit of swinging its head from side to side through a considerable arc as it swims, enables it to sample the smells in a much wider path through the water than would otherwise be possible.

Ralph E. Sheldon, working at the Marine Biological Laboratory in Woods Hole, Massachusetts, was the first to demonstrate scientifically that the sense of smell is important in guiding sharks to a meal. He found that smooth dogfish had no difficulty distinguishing a cloth-covered packet that contained crushed crabmeat from identical packets containing stones. When Sheldon plugged the sharks' nostrils with cotton wool so that a current of water could not enter the nasal sacs, they no longer homed in on the food packets, although they swam

quite close to them. To eliminate the possibility that the cotton wool plugs merely made the sharks uncomfortable and discouraged them from eating, Sheldon plugged the nostril on one side only of several dogfish. After a brief period of adjustment all but one of the animals readily located the packet containing food.

Several years ago, at the Lerner Marine Laboratory, scientists tested the sense of smell of lemon sharks *Negaprion brevirostris* by implanting electrodes in various parts of their brains (see box below). They concluded that the sharks were capable of detecting dilutions of one part of tuna juice to 25 million (10^6) parts of seawater. Albert Tester conducted laboratory tank experiments at Eniwetok in the 1960s and found that 'the response of blacktip and gray sharks to a standard extract increased under conditions of progressive starvation.' He found that starved blacktips were so sensitive to smells that they responded to concentrations of grouper flesh in dilutions of only one part in 10 billion (10^9) parts of seawater.

Sharks have excellent vision

REASSESSING THE MENTAL ABILITIES OF SHARKS

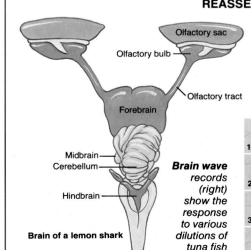

Brain of a lemon shark

0   20 mm

Brain wave records (right) show the response to various dilutions of tuna fish extract.

Sharks are often described as having 'primitive' brains – a claim that has not been borne out by recent studies. Brain size and complexity vary greatly from species to species, but some sharks obviously have considerable mental ability, and are capable of simple feats of learning. A comparison of brain to body-weight ratio (see p 21) shows that sharks compare well with many other vertebrates. The species with the largest brains in comparison to their body weights are the dusky shark and the scalloped hammerheads. In general, the active, fast-moving sharks have larger and more complex brains than the slower, bottom-dwelling species.

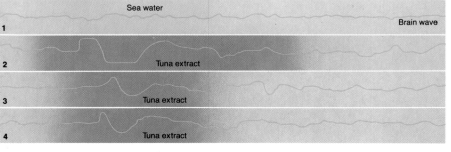

A great white shark 'sniffs' the surface of the water as it homes in on the source of a smell. Sharks trace odours by following the scent and always turning upstream when it is detected (see p.83).

The nostrils of the grey nurse shark *Eugomphodus taurus* are divided externally by a flap of skin. Water flows in through the openings on the outside of the nostrils, and out through the openings on the inner side of the flap. Once inside the shark's head the water passes through a funnel-shaped passage and into the nasal sacs, which are lined with sensitive cells.

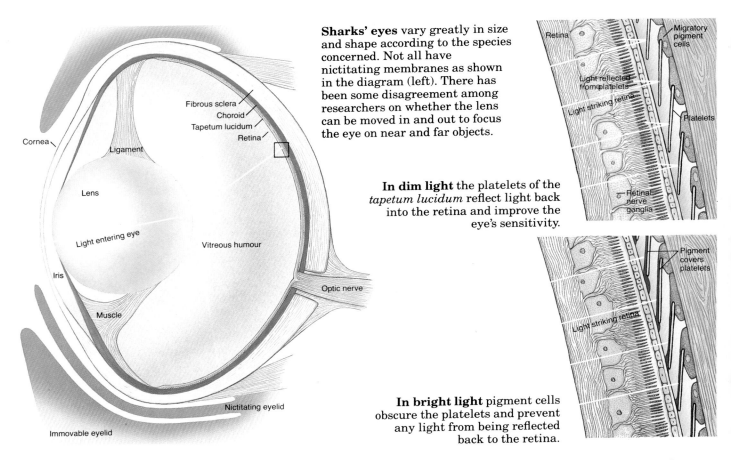

Sharks' eyes vary greatly in size and shape according to the species concerned. Not all have nictitating membranes as shown in the diagram (left). There has been some disagreement among researchers on whether the lens can be moved in and out to focus the eye on near and far objects.

In dim light the platelets of the *tapetum lucidum* reflect light back into the retina and improve the eye's sensitivity.

In bright light pigment cells obscure the platelets and prevent any light from being reflected back to the retina.

and experimental studies have now dispelled the widely-held notion that they have poor eyesight. As early as 1960, experiments were conducted in which large lemon sharks had discs placed over their eyes to blind them temporarily. The blinded sharks were then released into pens at the Lerner Marine Laboratory on Bimini in the Bahamas to compete for food with other sharks that had normal vision. The adult lemon sharks with unimpaired vision had no trouble homing in on and consuming a 113-kg (250-lb) chunk of blue marlin, while those temporarily blinded were unable to locate such an attractive feast. Other studies at this laboratory confirmed that sharks were attracted to bright objects while non-reflective, dull colours were generally avoided.

A shark's eye is a somewhat flattened version of the same eye that all vertebrates have, with an iris, lens, and retina, and three fluid-filled chambers contained within a tough envelope of cartilage, the sclera. In the shark, the aperture of the iris (pupil), varies in shape and, again, contrary to popular belief, may open and

close quite rapidly. In the nurse shark *Ginglymostoma cirratum*, for example, the pupil can dilate to its maximum size 24 to 30 seconds after entering the dark, and can constrict to its minimum size 5 to 13 seconds after emerging into the light again. In the brown shark *Carcharhinus plumbeus* maximum dilation and constriction take place in 40 and 25 seconds respectively. In every case the shark's pupil constricted in bright light more rapidly than it dilated in dim light or darkness.

The retina of the shark's eye contains many light-sensitive rod and cone cells, with there being many more rods than cones. The cone cells determine the extent of the animal's visual acuity and colour vision, and the large number of rod cells improve a shark's visual sensitivity, and therefore its ability to tell the difference between an object, especially a moving one, and its background in very dim light. In recent years experiments have shown that sharks have much greater visual ability than researchers first thought.

The sensitivity of the shark's eye in dim light is greatly improved

by a remarkable structure, the *tapetum lucidum,* a mirror-like reflecting layer that lies under the retina. The *tapetum* is made up of tiny platelets, silvered with guanine crystals, that can reflect incoming light back through the retina to restimulate the light-sensitive rods and cones. The *tapetum* therefore enables sharks that feed at night – as most species that live in the open seas do – to make the most of the small amount of light that is available.

The *tapetum,* however, has an even more remarkable ability. For those sharks that also feed during the day, bright light must be prevented from reaching the delicate cells of the retina. This is achieved not only by reducing the size of the opening in the pupil, but also by a curtain of pigment that temporarily screens each of the tiny platelets of the *tapetum.* Each platelet is covered by cells that contain pigment, and black granules temporarily fill the cells in bright light, thus preventing light from being reflected back into the retina. Conversely, as the shark's eye becomes accustomed to the dark, these black granules

The bulbous ampullae of Lorenzini are shown in this cross-section connected to canals that lead to the skin surface, and to nerves. The diagram of the shark's head shows the locations of skin pores leading to both the ampullae of Lorenzini and the lateralis system.

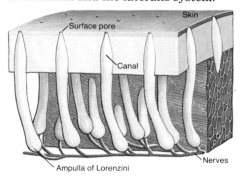

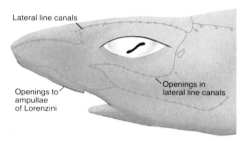

Skin pores show up clearly in this view of the head of a grey nurse shark. They lead to the ampullae of Lorenzini – sensory organs which enable sharks to detect very weak electric fields, and therefore any prey that may be buried under sand on the sea floor.

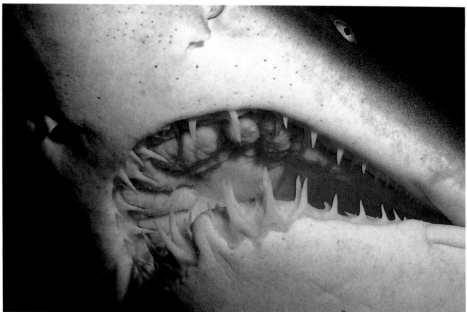

withdraw to the base of the cells to expose the reflective surface of the platelets again.

The lens of a shark's eye is nearly spherical and rigid, unlike the ellipse-shaped, elastic lens that mammals have. Some sharks can focus on an object by moving the lens towards or away from the retina, much as a camera can be focused. The movement of the lens is controlled by a muscle – the protractor lentis.

It has now been conclusively demonstrated that sharks can locate their prey in the open sea by detecting the minute electric fields they generate. One of the first to suspect that sharks may be sensitive to tiny electric currents was Sven Dijgraaf, working at the University of Utrecht more than 45 years ago. He noticed that the blindfolded, small-spotted catsharks *Scyliorhinus canicula* he was using in behavioural experiments would turn rapidly away from a rusty steel wire located a few centimetres from their heads. However, when a glass rod was held at the same distance it did not produce the same response. Dijgraaf assumed that his sharks were responding to electric currents

A researcher uses an electrode at the end of a pole to test the electrical sense of this shark. The extraordinary sensitivity of this sense has only been fully appreciated in recent years.

generated at the surface of the rusty wire. But what receptors did the sharks employ for doing this? A clue was provided in the early 1960s by R.W. Murray who demonstrated that the ampullae of Lorenzini on the snouts of sharks responded to

weak electric fields. When Dijgraaf and one of his bright graduate students, A.J. Kalmijn, severed the nerves to these ampullae the sharks were no longer able to detect electric currents.

It remained for Kalmijn to show that a shark may employ these unique electroreceptors to locate prey. In a classic series of experiments he demonstrated that a small-spotted catshark can locate a live flounder buried beneath the sand in a laboratory aquarium. It did this by detecting minute electric pulses generated by the resting flounder. The shark was even able to find the flounder when it was placed in an agar chamber that permitted seawater and electric pulses to pass through. When the agar chamber was covered with a plastic film, electric pulses from the flounder were blocked, and the shark could not locate it. Finally, when a pair of electrodes were buried in the sand to simulate the flounder, Kalmijn's sharks attacked the electrodes.

During the summer of 1976, Kalmijn and K.J. Rose continued their work in the open sea off Cape Cod, Massachusetts. They succeeded in getting the nocturnal,

The lateralis system alerts sharks to low-frequency vibrations, such as distant disturbances in the water made by struggling fish, or the erratic movements of a human swimmer. This sensitivity to vibration is shared by most fishes and some aquatic amphibians. There are still many unanswered questions about the role that this system plays in the everyday lives of sharks, in spite of the fact that it has recently been the subject of considerable research.

Lateral line canals

Longitudinal fibres
Supporting cells
Gelatinous dome

Neuromasts

Hairlike projections

Canal

Tube

Tube

Skin opening

Tube

Skin

Skin opening

Tube

Skin opening

Skin opening

Tests have shown that the lateralis system is extremely sensitive. Blinded sharks are even able to detect the wall of a large tank without touching it, apparently by sensing water waves reflected from the surface.

bottom-feeding dusky smoothhound *Mustelus canis* to 'home in' on live electrodes which were giving off weak electric (DC) pulses. The sharks ignored 'dead' electrodes and other stimuli. After observing this identical behaviour in hundreds of other dusky smoothhounds in their natural environment, Kalmijn and Rose concluded that the sharks can detect their prey at close range using only their electric sense.

So responsive are sharks to a very weak electric field that Kalmijn believes they may also be sensitive to the earth's magnetic field and can use this ability to navigate. The dusky smoothhound regularly migrates southward from the waters off Cape Cod, Massachusetts, for the winter months, and is apparently endowed with a good sense of direction. The ampullae of Lorenzini may

therefore provide the sharks with a very accurate internal electromagnetic compass.

During his experiments Kalmijn found that 'within the frequency range of direct current up to about eight hertz (cycles per second), sharks respond to fields of voltage gradients as low as a hundred-millionth of a volt per centimetre. That would be equivalent to the field of a flashlight battery connected to electrodes spaced 1000 miles [1600 km] apart in the ocean'! This remarkable electrical sensitivity is greater than that possessed by any other animal investigated so far.

Sharks have an acute vibration sense called the lateralis system which is located in their heads and bodies. It consists of several small canals that open at intervals to the surface through tubes to pores in

the skin. Each canal is filled with seawater and contains clusters of sensory cells, called neuromasts, on its inner surface. From each neuromast several hairlike projections, enclosed in a gelatinous dome, stick out into the interior of the fluid-filled canal. Vibrations that reach the shark's head and body are transmitted to the neuromasts and cause the hairlike projections to move very slightly, thereby triggering a nerve impulse to the brain.

More than half a century ago George H. Parker of Harvard University found that a dogfish that could not see or hear was still able to detect disturbances in the water as long as its lateralis system was undamaged. When he severed the nerves of the lateralis system, the shark ceased to respond.

In addition to the lateralis

Shark's ears have a number of primitive features which are still little understood. Both the endolymphatic duct and the macula neglecta have disappeared, or have been reduced, during the evolution of vertebrate ears. Shark ears have been particularly well studied because they have long been a favourite subject for dissection to test the skill of aspiring medical students — they contain the same basic structures that are to be found in the ears of all other vertebrates.

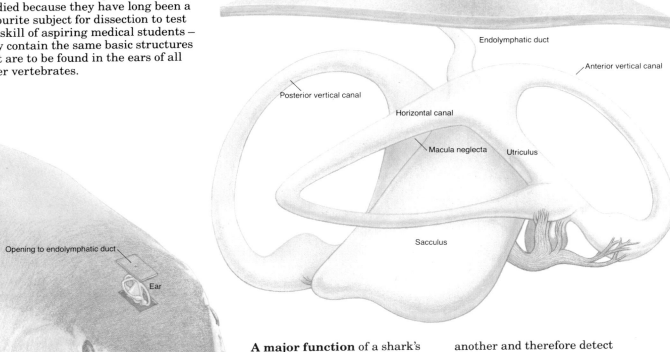

Skin — Opening in head

Endolymphatic duct

Anterior vertical canal

Posterior vertical canal

Horizontal canal

Macula neglecta

Utriculus

Sacculus

Opening to endolymphatic duct

Ear

A major function of a shark's ears, and indeed any fish's ears, is to provide the creature with information about its position and speed. In a watery world there are few visual clues to help with this task. The three semicircular canals are at right angles to one another and therefore detect movement in any one of the three spatial planes. Information of acceleration and deceleration is provided by the otoliths when they move, and liquid surging in the canals tells the shark's brain about the extent of a turn.

system, sharks also detect vibrations with their ears. The ears of a shark — one per side — are enclosed in its cartilaginous brain case, and are made up of the same basic components that are found in all vertebrate ears — three semicircular canals, the utriculus, the sacculus, and their related structures.

Recent evidence suggests that sharks can pick up low frequency sounds from great distances. Work by A.N. Popper and R.R. Fay in the 1970s showed that the shark's ear did this by detecting the movement of water particles, which carry vibrations, rather than from changes in sound pressure. The three fluid-filled semicircular canals in the shark's ear detect changes in the speed and direction of the shark's movement. Other organs in the sacculus and macula neglecta

contain otoliths — tiny granules of calcium carbonate — which move when the shark changes position, and send signals to the brain to tell it about the animal's attitude in the water. The macula neglecta is especially sensitive to vertical movements, and it can also pick up vibrations from the roof of the animal's skull.

Information from both the ears and the lateralis system may be used to locate the source of a sound quickly and accurately. It appears that the macula neglecta may be particularly important for doing this, a theory that is borne out by measurements made by J.T. Corwin. He examined the ears of a number of species of sharks and found that the macula neglecta was most elaborate in the grey reef shark, an active predator, less elaborate in the dusky smoothhound and nurse

shark, and still less elaborate in the rays, where directional hearing is of least importance for locating food.

Another curious feature of shark ears is a tiny duct — called the endolymphatic duct — which extends from the sacculus of both ears to small pores on the shark's head. The exact purpose of these ducts is as yet unknown.

Sharks also have one other sensory system which is still little understood. This consists of a large number of 'pit organs' scattered along the animals' bodies and around their lower jaws. These are usually protected by modified denticles, and closely resemble the taste buds of higher vertebrates. Opinion is divided on whether the organs have a taste function, whether they pick up vibrations, or even possibly sense movements in the shark's body.

INHABITANTS OF ANCIENT SEAS

There were sharks in the oceans of earth long before the first animals had begun to colonise the land surface. Their history stretches back for at least 400 million years, a vast period of time which makes the 2-million-year-old history of humans (genus* Homo*) seem insignificant by comparison.

The planet earth is at least 4500 million years old. The earliest signs of life date back 3000 million years, and by 600 million years ago quite complex animals and plants existed.

The story of the backboned (vertebrate) animals – the fishes and other forms that eventually led to humans – began about 600 to 500 million years ago. The first definite fossil of a creature with a stiffening notochord along its back was a Cambrian animal called *Pikaea*, which was discovered in 560-million-year-old rocks in Canada. The oldest complete fish fossil so far found came from central Australia, and was a jawless, armoured fish called *Arandaspis*, which dates from the early Ordovician period. This small filter-feeding or mud-grubbing fish lived in shallow coastal waters over 480 million years ago.

In the next 100 million years all the major fish groups that are found on earth today evolved, although their origins and relationships to one another are still uncertain. In the Silurian period – some 400 million years ago – the jawless fishes were the most numerous, although primitive bony and cartilaginous jawed fishes had begun to appear at about that time. Fossils of the so-called 'spiny sharks' – which had large fin spines – and of the earliest true sharks have also been found in Silurian rocks. By the Devonian period – about 30 million years later – fishes had diversified and spread into all parts of the world. By the end of the Devonian, some 25 million years further on, certain fishes had moved onto land. From these first amphibians evolved the four-legged animals, including reptiles such as dinosaurs, and eventually mammals.

The early evolutionary history of sharks and shark-like fishes is still poorly understood. Until recently scientists thought that there were no shark fossils to speak of in rocks older than those from the Middle Devonian. Now, however, it is certain that sharks did not appear suddenly at that time – it was just that researchers were looking for the wrong kind of evidence. Microscopic examination of ancient sediments has revealed fossilised remains of sharks which may push their origins back at least 50 million years further.

Traditionally, sharks have been regarded as primitive vertebrates – so-called 'living fossils' – but recent work suggests that they are highly specialised. Their complex biology ranks them with birds and mammals as highly evolved

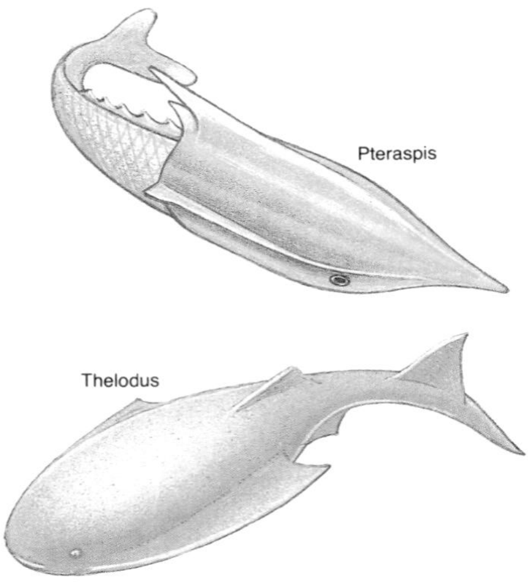
Pteraspis
Thelodus

Two of the earliest backboned animals on earth. These ancient fishes, which lived in seas that covered the earth between 350 and 440 million years ago, had primitive, sucking mouths.

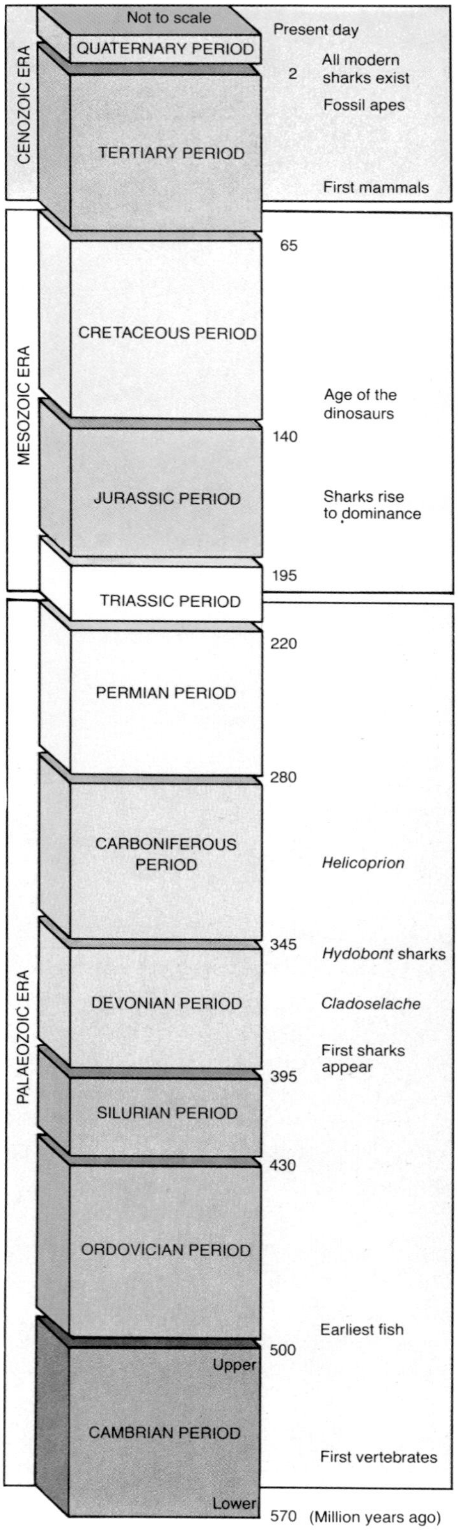
Not to scale
Present day
CENOZOIC ERA
QUATERNARY PERIOD
2
All modern sharks exist
Fossil apes
TERTIARY PERIOD
First mammals
65
MESOZOIC ERA
CRETACEOUS PERIOD
Age of the dinosaurs
140
JURASSIC PERIOD
Sharks rise to dominance
195
TRIASSIC PERIOD
220
PERMIAN PERIOD
280
CARBONIFEROUS PERIOD
Helicoprion
345
Hydobont sharks
PALAEOZOIC ERA
DEVONIAN PERIOD
Cladoselache
First sharks appear
395
SILURIAN PERIOD
430
ORDOVICIAN PERIOD
Earliest fish
500
Upper
CAMBRIAN PERIOD
First vertebrates
Lower
570 (Million years ago)

This chart shows the subdivisions of geological time, from the Cambrian period to the present day. The terms Lower and Upper are used to describe the beginning and end of a period.

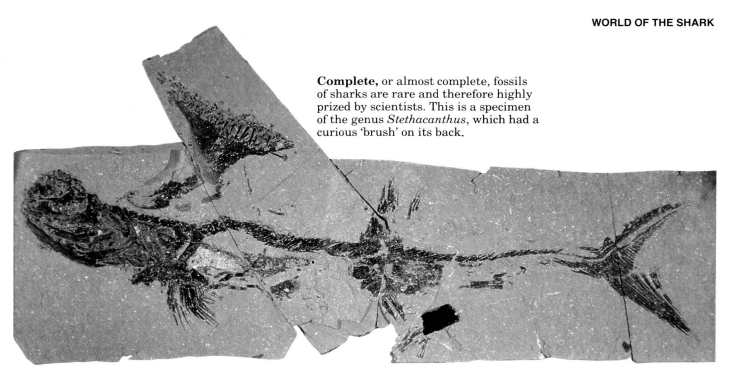

Complete, or almost complete, fossils of sharks are rare and therefore highly prized by scientists. This is a specimen of the genus *Stethacanthus*, which had a curious 'brush' on its back.

creatures. Even Devonian sharks, which were once considered to be the most primitive, are now thought to have been relatively specialised – with a longer evolutionary past than was once supposed.

The evolution of sharks remains unclear partly because it has been difficult to analyse particular characters – such as the shapes of braincases or fin structures – of modern sharks, which are an extremely diverse group, adapted to many habits and habitats. Any analysis of fossil sharks has had to rely almost solely on hard parts, such as teeth. Other features such as fins are only rarely preserved. Much more needs to be known about ancient sharks before any assessment can be made as to which characters were 'primitive', and which were 'advanced'.

All that is known about ancient sharks and their evolution has been gleaned from the fossil record. Fossil remains of sharks have been known for many centuries, although their true nature was not always recognised. Until the 17th century many scholars regarded such fossils as sports of nature, and thought that fossilised shark's teeth were bird's or snake's tongues. Some large fossil teeth, called glossopetrae (literally tongue-stones), were used as amulets to ward off evil and to protect against poisoning. It was not until about 150 years ago that the study of fossils, palaeontology, became a science, and the ancient remains of plants and animals were systematically classified.

Sharks are rarely found as complete fossils because their skeletons are made of cartilage. Normally only the hard parts, such as teeth, scales and fin spines, are found. However, under certain special conditions, complete fossil sharks are preserved, and these provide scientists with vital information. One such deposit, found last century in Upper Devonian Cleveland shales from the USA, yielded entire shark carcases which had been preserved in a bacteria-free environment so that even muscle and kidney tissue could be examined in the rock.

The very earliest signs of sharks are minute fossil scales and teeth which are found in rocks from the late Silurian to early Devonian

Fossil shark teeth embedded in a piece of Miocene limestone from Victoria in Australia. Detailed examination of such remains has enabled scientists to piece together the complex story of how life evolved on earth.

period (around 400 million years ago). It becomes more and more difficult, however, to identify shark scales in older rocks because they closely resemble those from jawless fishes called thelodonts, which lived at the same time. Only microscopic differences separate shark and thelodont scales, and the two kinds seem to become more and more alike the further one goes back.

A similar problem exists with ancient shark teeth, which did not seem to be present in rocks older than those from the mid-Devonian. It now seems that the reason for this was that scientists were not looking in the right places, and that early shark teeth were often very small. In 1986 teeth were found in Lower Devonian rocks from Spain which belonged to a group of sharks called xenacanthids.

What, then, were the origins of the shark-like fishes? One possibility is that different cartilaginous fishes (the group which includes sharks, skates, rays and chimaeras) evolved from placoderms (now extinct, bony-plated jawed fishes). Alternatively the placoderms and the cartilagenous fishes might all have shared a common ancestor at some time in the early Silurian period, some 430 million years ago. A third possibility is that the cartilaginous fishes and a group of primitive jawless fishes, such as thelodonts, both had a common ancestor. Thelodonts (now extinct) had a skeleton made of cartilage and were covered with scales, which may have lined their mouths as well. They also had paired fins and eight pairs of branchial structures for supporting their gills. One region where thelodonts evolved rapidly was around Tuva, in Siberia, which is also where the oldest known shark scales have been found.

The common ancestor of sharks and thelodonts (a fossil of which is yet to be found) would probably have been a small fish with a long slender body, one dorsal fin, no fin spines, paired pectoral fins and at least seven pairs of gill supports. Its mouth would have been either at the front of its body, or slightly beneath it, and the creature would have been covered with small scales which varied in size and shape according to their position on its

ANCIENT ANCESTORS OF MODERN SHARKS

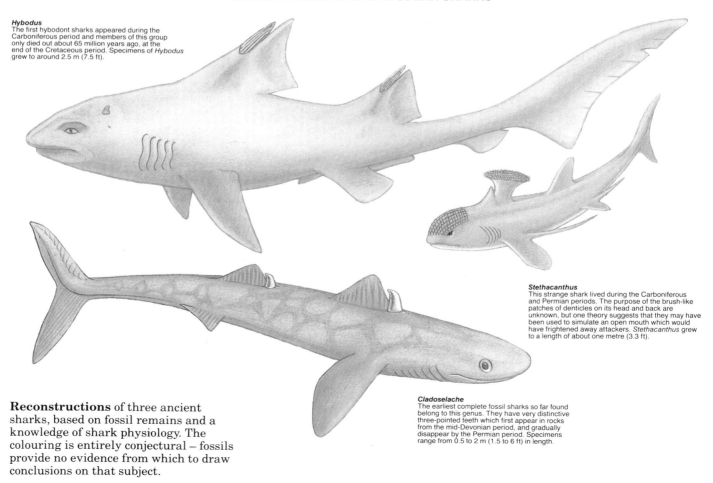

Hybodus
The first hybodont sharks appeared during the Carboniferous period and members of this group only died out about 65 million years ago, at the end of the Cretaceous period. Specimens of *Hybodus* grew to around 2.5 m (7.5 ft).

Stethacanthus
This strange shark lived during the Carboniferous and Permian periods. The purpose of the brush-like patches of denticles on its head and back are unknown, but one theory suggests that they may have been used to simulate an open mouth which would have frightened away attackers. *Stethacanthus* grew to a length of about one metre (3.3 ft).

Cladoselache
The earliest complete fossil sharks so far found belong to this genus. They have very distinctive three-pointed teeth which first appear in rocks from the mid-Devonian period, and gradually disappear by the Permian period. Specimens range from 0.5 to 2 m (1.5 to 6 ft) in length.

Reconstructions of three ancient sharks, based on fossil remains and a knowledge of shark physiology. The colouring is entirely conjectural – fossils provide no evidence from which to draw conclusions on that subject.

body. Scales would also probably have lined its mouth.

The story of how sharks developed from these ancestral forms is complex and confused. Scientists do not have a set of fossil remains which show a smooth transition from species to species, connecting ancient forms to their modern descendants. Instead they have some isolated pieces from an immensely complex jigsaw. Some fossils are of evolutionary experiments which led nowhere and eventually became extinct. Others may have features which seem to explain gaps in our knowledge, but also have other features which raise more questions. Some new discoveries provide little information at present, but may

become vital when other pieces of the jigsaw are in place. Many theories exist to explain the evidence that is available, and these are constantly being modified as new information comes to light.

Although there is evidence of earlier sharks, the first complete fossils of shark-like fishes have been discovered in mid-Devonian rocks. Most frequently found are members of the genus *Cladoselache*, streamlined fish that grew to a length of about two metres (6 ft). Fortunately, complete specimens of *Cladoselache* have been preserved in the remarkable Cleveland shales, so quite a lot is known about them. They had five pairs of gill slits, a fin spine and all the same fins as modern sharks, except for an anal

fin. These spines, which become more common and elaborate in later sharks, and which still persist in some species today, were positioned in front of the dorsal fins and acted as cutwaters. *Cladoselache* had distinctive teeth with a large central cusp flanked by several smaller points, and apparently they lived on small fish – the remains of which have been found in the stomachs of fossilised specimens. These sharks are not now considered to be the main line leading to the modern species.

One important feature, however, was missing from *Cladoselache*, and most of its relatives in Devonian seas – claspers. These copulatory devices, which all modern sharks have, are only found in the

Shark teeth, together with scales and dorsal spines, are the three parts of sharks most often preserved in rocks. Both crushing teeth, for eating shellfish, and pointed teeth, for eating flesh, are included in this collection.

A chart showing the way in which the various groups of modern sharks might have evolved. Diagrams such as this are based on a study of fossil remains to see how ancient, as well as modern species are related. Fine details of the evolutionary tree change as new information becomes available.

THE SHARK'S FAMILY TREE

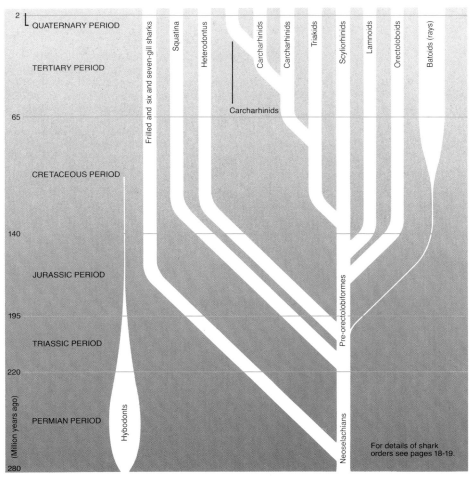

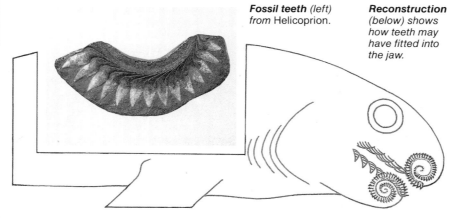
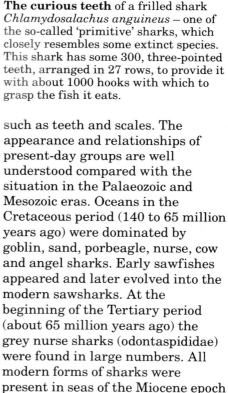

The curious teeth of a frilled shark *Chlamydosalachus anguineus* – one of the so-called 'primitive' sharks, which closely resembles some extinct species. This shark has some 300, three-pointed teeth, arranged in 27 rows, to provide it with about 1000 hooks with which to grasp the fish it eats.

stethacanthids and a few other groups of ancient sharks that lived at the same time as *Cladoselache*.

In the Devonian period a new group of sharks became common – the hybodonts. These became the dominant group during the Mesozoic era (from about 220 to 65 million years ago), when nearly all other shark groups died out. Although they were not as streamlined as modern oceanic sharks, and were probably not as accomplished as swimmers, they were nevertheless an advance on *Cladoselache*. A typical genus of this group is *Hybodus* which was found all around the world from the Triassic period to the mid-Cretaceous. This shark grew to at least 2.5 m (7.5 ft) in length, and had a blunt snout with an elongated body. The arrangement of its teeth – with sharp pointed ones at the front and blunt ones at the back – seems to suggest that it caught swimming prey as well as eating shellfish, such as molluscs.

Although sharks and shark-like fishes have a long history, the modern sharks (neoselachians) did not rise to dominance until after the Jurassic period, when, for some reason that is not yet clear, many of the more ancient forms had become extinct. Some Jurassic sharks are closely related to modern sharks, and this gives many present-day shark families histories which stretch back for 135 million years or more. The skates and rays, another group of cartilaginous fishes, also appeared in the mid-Jurassic, but they did not really come into their own until the Tertiary period, between 65 and 2 million years ago, when they were able to exploit a dramatic rise in the numbers of bivalve shellfish in the oceans.

The fossil record of modern sharks is fairly good, but again it normally consists only of hard parts such as teeth and scales. The appearance and relationships of present-day groups are well understood compared with the situation in the Palaeozoic and Mesozoic eras. Oceans in the Cretaceous period (140 to 65 million years ago) were dominated by goblin, sand, porbeagle, nurse, cow and angel sharks. Early sawfishes appeared and later evolved into the modern sawsharks. At the beginning of the Tertiary period (about 65 million years ago) the grey nurse sharks (odontaspididae) were found in large numbers. All modern forms of sharks were present in seas of the Miocene epoch (25 to 5 million years ago), including the giant *Carcharodon megalodon*, perhaps the most awesome of all sharks, now extinct.

Many researchers are working on the problems presented by ancient sharks, and recent discoveries have included some of the oldest shark fossils found. Among international research has been work in Germany which has concentrated on examining the fine structure of shark teeth and scales, and has revealed that the outer enamel on many teeth has three layers – a feature which can be used to identify certain groups of sharks. Other work, in the USA, has been delving into the nature of the supposedly primitive modern species – the frilled shark *Chlamydoselachus anguineus*, the Port Jackson shark *Heterodontus portusjacksoni* and the sixgill and sevengill sharks (family hexanchidae) – in an attempt to understand their relationships to other modern sharks, and to uncover any links with earlier fossil forms.

One particularly important recent discovery has been the recognition of a new class of fossil sharks called iniopterygians. Remains of this strange group have been found in Upper Carboniferous rocks from North America which are 300 million years old. Examination has shown that these creatures are a 'missing link', and that they combined certain characteristics of both elasmobranchs (sharks, skates and rays) and holocephalians (chimaeras or rattails). The life styles of these fishes remain uncertain, although their complex tooth plates seem to suggest that they lived on shellfish.

There has been an upsurge of interest in fossil sharks in recent years as more information has become available from new fossil discoveries, especially in the southern continents, and from a microscopic examination of existing fossil remains. In many cases these discoveries have helped scientists to understand some of the finds made last century. There are still few definite answers about the origins and relationships of all the known fossil, and some modern, sharks but progress is being made.

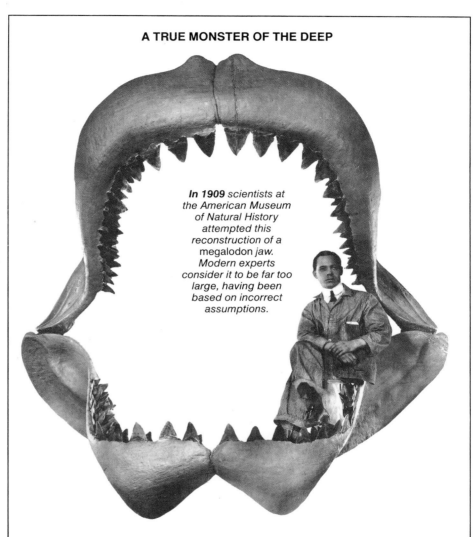

A TRUE MONSTER OF THE DEEP

In 1909 scientists at the American Museum of Natural History attempted this reconstruction of a megalodon *jaw. Modern experts consider it to be far too large, having been based on incorrect assumptions.*

Some 12 000 years ago, at about the time when mammoths became extinct, humans who ventured into the sea faced the possibility of meeting the largest flesh-eating shark that has ever lived – a close relative of the great white.

Estimates vary for the length of *Carcharodon megalodon* – as this monster has been named – but the most realistic seem to agree on a figure of between 12 and 15 m (40 and 50 ft), with a weight of around 20 tonnes.

Teeth from this giant shark – some measuring up to 152 mm (6 in) in length, and weighing 340 g (12 oz) – are commonly found in Palaeocene sediments from around the world. Many have also been dredged from the sea floor (see p 70).

Nobody can be sure exactly when *megalodon* became extinct, and some optimists have even claimed that isolated specimens may still exist in the ocean depths today. Most scientists, however, dismiss this possibility out of hand.

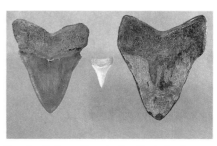

Two teeth from the giant extinct shark Carcharodon megalodon, *compared with that of a great white. The fossil teeth are large – probably as large as any found – and came from a shark that must have been at least 12 m (40 ft) long. The great white to which the single tooth belonged was around 3 m (9.8 ft) long. The enamel height on the largest* megalodon *tooth ever found is 115 mm (4.53 in), compared with a maximum for a great white of around 70 mm (2.8 in).*

SHARKS IN THE WILD

It is difficult to study sharks in the ocean, particularly the big predators, because they are almost impossible to watch for any length of time. Before techniques for tagging large sharks were developed, very little was known of their movements and migrations.

In the 1950s various tags were developed which could be inserted into fishes without removing them from the water. This is especially important with sharks because their organs are supported by membranes which tear when subjected to the full force of gravity.

The techniques for tagging large fishes were developed at Woods Hole Oceanographic Institute, Massachusetts, in the USA. Fish were caught with conventional sport fishing gear and a long pole was used to prod a sharpened tag into the shoulder of each animal before it was released.

Because this method proved to be most effective for catching and tagging large sharks, scientists took advantage of a growing interest in sport fishing, and issued tags to recreational anglers. This voluntary system has been so successful that many fisheries' programmes throughout the world now rely completely on the assistance of thousands of dedicated anglers.

By far the largest co-operative

A tagged copper shark *Carcharhinus brachyurus* about to be released off the New South Wales coast. The white tag can clearly be seen projecting from beneath the fish's dorsal fin.

shark tagging programme in the world is conducted from the National Marine Fisheries Service laboratories in Narragansett, Rhode Island, USA. The programme has been running for over 20 years, and in that time over 50 000 sharks have been tagged. Recreational anglers have carried out most of those taggings, the balance being done by scientists on board research and commercial vessels. The results have been outstanding and have helped in an understanding of the growth, movements and population structures of a number of species of large sharks.

Probably the most complete picture of migration patterns has come from the tagging of over 30 000 blue sharks *Prionace glauca*. Many transatlantic crossings have been recorded, by sharks starting from both sides of the Atlantic, as well as other long distance movements throughout the region. In 1985 the first blue shark tagged in the western Atlantic was recaptured over the equator in the southern hemisphere. Until then, it was thought that blue sharks in the north Atlantic were part of one stock, distinct from those in the south Atlantic.

Long distance movements have also been recorded by shortfin mako, tiger and sandbar sharks. The greatest distance recorded for a tagged mako was 2700 km (1690 miles) from Virginia, USA, to the West Indies. The fastest sustained trip for this species was made by a fish which, in just three months, swam over 2413 km (1500 miles) at a rate of 28 km (17.6 miles) per day. Of course, this is the minimum rate of travel for a straight-line journey. In reality, the shark would have covered a far greater distance.

A tagged tiger shark swam 2965 km (1853 miles) from New York to Costa Rica, while many sandbar sharks have been tagged off the eastern United States and

FITTING THE PUZZLE TOGETHER

The results of tagging studies are slow to realise. Only a small proportion of the fish tagged are ever recaptured, so many tens of thousands must be released before any pattern of behaviour will begin to emerge. In 1985 over 7000 sharks were released in the United States' programme, and only 256 sharks were captured with tags in them. The 256 tags were returned from 15 countries around the world, as far apart as Japan, Spain, Puerto Rico, Canada and Korea. Most of the recaptured fish (48 per cent) were caught by commercial long-line fishermen, with 36 per cent being caught by amateur anglers fishing with a conventional rod and line.

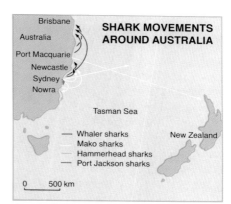

A mako holds the Australian record.

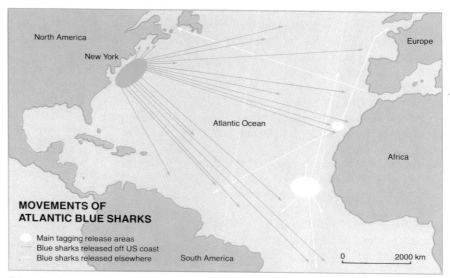

Blue sharks are the greatest travellers of all – some swimming 6000 km (3800 miles).

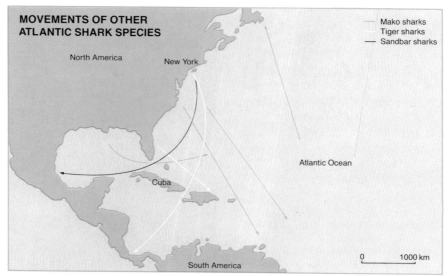

Sharks of these three species regularly swim 2700 km (1700 miles) or more.

recovered in Mexican waters, 2900 km (1800 miles) away.

Another co-operative gamefish tagging programme, modelled on the successful American example, started in Australia in 1973. This programme, with its headquarters at the New South Wales Department of Agriculture's Fisheries Research Institute in Sydney, seeks the co-operation of anglers to tag a wide range of gamefish including marlin, tuna and the so-called game sharks.

Over 40 000 fish have been tagged by volunteer anglers since the programme started, including around 3000 sharks. The whaler sharks (family Carcharhinidae) are the most frequently tagged.

Nearly two per cent of all sharks tagged in Australian waters have been recaptured. Most of those recovered were whalers, with makos second. All of the whaler sharks recaptured were found to the north of their release points. This is interesting because most sharks of this group tagged off the eastern United States migrated to the south. Sharks in both hemispheres seem to move towards the equator.

The longest movement recorded so far was by a mako shark which crossed the Tasman Sea from Australia to New Zealand, a distance of 2050 km (1285 miles). The fastest rate of travel was only 3.3 km (2 miles) a day for a whaler shark at liberty for 86 days. This is much less than the United States record, but similar results are expected as more tagged sharks are released each year.

Apart from providing information about migration

Scientists grapple with a small shark on the deck of a research vessel as they attempt to insert a tag. Even small sharks like this are surprisingly strong and can only be held down with some effort. The creature's razor sharp teeth could inflict considerable damage if it managed to break free.

RECORD TRAVELLERS

Dozens of blue sharks Prionace glauca *mill about in the water beneath a boat.*

The most travelled species of shark (and one which has been tagged in large numbers) is without doubt the blue shark. Tagging has shown that the species wanders throughout the north Atlantic Ocean.

Many sharks are popularly thought to be territorial, but tagging shows that most species do move around to some extent. Perhaps the least travelled group are the nurse sharks.

The record distance travelled by a shark is held by a blue which swam 5980 km (3740 miles) from New York State, where it was released, to Brazil.

The time record between tagging and recapture is held by two sandbar sharks. Both were tagged in the same week in 1965 off the eastern coast of the USA. Incredibly, they were both recaptured in the same week, 19.7 years later, over 1600 km (1000 miles) from where they were released, but only 160 km (100 miles) apart!

patterns, tagging programmes can also help in studying other aspects of the biology of sharks, such as growth rates and population structures.

It is important to know how sharks (and other fishes) grow and age so that replacement rates can be estimated. Only then is it possible to say how many sharks can be taken by fishermen without seriously damaging stocks. Because tagged sharks are actually measured when they are released and recaptured, often some years later, it has been possible to show that some species grow only a few centimetres a year. This, together with the fact that often very few young are produced in a season, means that such species can easily be over-exploited.

Tagging has also shown that some sharks keep close to the coast all of their lives, while others, such as blue sharks, may cross oceans and enter the fishing zones of other nations. This information is vital when 'ownership' of fish stocks are discussed internationally. Studies have also revealed that males and females of some shark species are often segregated for much of their lives, only mixing during the breeding season.

At present tagging can only show where a fish was released and recaptured, while the history of its actual day-to-day movements and activities remain a mystery. However, a new type of tag may help to provide this information. Known as the sonic tag, it is a small, battery-powered transmitter attached to a normal tag head. The

STUDIES THAT SHOW HOW SHARKS AGE AND GROW

Sharks, like other fishes, continue to grow throughout their lives. Early growth is usually fairly rapid in most species, tapering off after maturity, and eventually becoming very slow, but nevertheless still perceptible, throughout later life.

It is fairly easy to calculate the age of bony fishes, such as trout or perch, because scales and bones have annual growth rings. Sharks, however, have skeletons made up mostly of cartilage and only the core of the skeleton, the vertebrae, show any growth rings which can be used to estimate age.

To age a shark several vertebrae are cut from the animal's body and the central part is removed and treated in some way to expose the growth rings. This can be done by etching the bone with mild acid and silver nitrate, scorching or polishing. When it has been treated, the surface of the vertebra will usually show concentric rings, like those on a cut tree trunk – and in much the same way, the age of the shark can be estimated by counting these rings. In practice, there are usually many more rings than there are years that the shark has been alive, and

this is where tagging studies can assist scientists in deciding which rings to count.

If a shark is recaptured one, two or more years after being tagged, then its vertebrae can be used to count likely annual rings. This method is especially useful if very small sharks of known age (say less than one year old) are tagged, and subsequently recaptured several years later. The real annual rings can then be determined and these results applied to all other samples. For this reason, scientists are always excited when they receive the whole carcass of a recaptured shark that has been at liberty for more than one year.

Ageing studies of sharks have shown that some species are very slow growing, while others are much faster. The average growth rate of sandbar sharks *Carcharhinus plumbeus,* for example, has been found to be about 43 mm (1.7 in) per year, whereas the shortfin mako apparently grows at the much faster rate of about 280 mm (11 in) per year. Blue sharks grow even faster, at about 320 mm (12.6 in) per year. Individual sharks of the same species may also have very different growth rates (see below).

Growth rings in the vertebra of a female spot-tail shark (above).

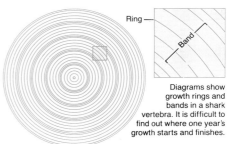

Diagrams show growth rings and bands in a shark vertebra. It is difficult to find out where one year's growth starts and finishes.

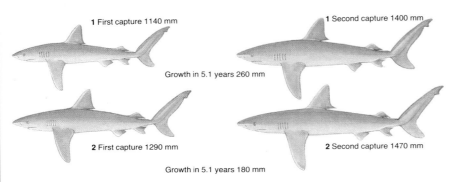

1 First capture 1140 mm

Growth in 5.1 years 260 mm

2 First capture 1290 mm

Growth in 5.1 years 180 mm

1 Second capture 1400 mm

2 Second capture 1470 mm

Two Galapagos sharks Carcharhinus galapagensis *tagged and recaptured off South Africa. Both had been at liberty for*

5.1 years, and had grown at quite different rates. The growth rate of the first shark had almost doubled that of the second.

Various chemical processes *can be used to accentuate growth patterns in the vertebrae (spinal bones) of sharks. The patterns take the form of bands, each made up of a number of closely-spaced rings. Experiments are now being carried out to inject captured sharks with chemicals that are quickly deposited in bone. When sharks that have been treated in this way are recaptured, their vertebrae can be examined to show up the band of injected chemical, and any subsequent growth rings that can be counted will therefore have been formed over a known period of time.*

device emits sonic pulses which vary with depth, and these can be tracked and recorded using a receiver on a boat.

The system has been used successfully on billfish and tuna, but only recently on sharks. One such experiment involved tracking a large white shark for two and a half days. In that time, it cruised at an average speed of 3.2 km/h (2 mph) and covered about 190 km (120 miles).

One possible problem with this type of experiment is that sharks may not behave normally for a while after being hooked and tagged. It has not been possible to track sharks for more than a few days, and it is during this period that their behaviour is most likely to be abnormal.

A number of different types of tags were used to mark sharks in the past, including internal plastic discs, plastic clips attached to dorsal fins, and even freeze-branding with very cold implements. Most modern shark tags resemble small arrows or darts, made from plastic, nylon and stainless steel. These are inserted into the shoulder muscle with a short jab using a tag pole.

Few sharks die after being tagged and released, and the process does not appear to upset them for long. There are many cases of tagged sharks being recaptured the same day that they were released, and even of freshly tagged sharks doggedly following the boat which released them!

Once the tag is embedded and fully healed, it is likely that it will stay in place for several years. Hooks are sometimes seen in the mouths or stomach linings of recaptured sharks, but because very few are found, it seems that they corrode and fall out soon after the fish is released. Only rarely do wounds become ulcerated.

The two main international tagging programmes in which sharks are included are detailed in the box (right). Programmes involving anglers also exist in Great Britain, South Africa, Canada, and other parts of the United States. The main species studied are the large predators, including the blue, mako, tiger, hammerhead and white sharks, and various species of the family Carcharhinidae. Most of these are distributed unevenly around the world and scientists hope that tagging will help to show if they are interrelated.

A fisherman uses a tagging pole to insert a tag in a small shark that has been hooked and is being held alongside the boat by another angler. This experience apparently causes the sharks no lasting harm.

STATE FISHERIES
N.S.W.
Gamefish
TAGGING
The Game Fishing Association of Australia,
the Australian National Sportsfishing Association
and New South Wales State Fisheries
Boat:
1st tag: S 32016
Skipper:
2nd tag:
Date: Day Month 198 Fighting Time: Hrs Mins
Locality:
Species:
Measured length: Estimated
Actual weight: Estimated
Remarks:
Name:
Address:
Club: P/code
Is this your first tag and release? Yes No
Do you need more tags? Yes No
Please complete and mail this card TODAY
SPONSORED

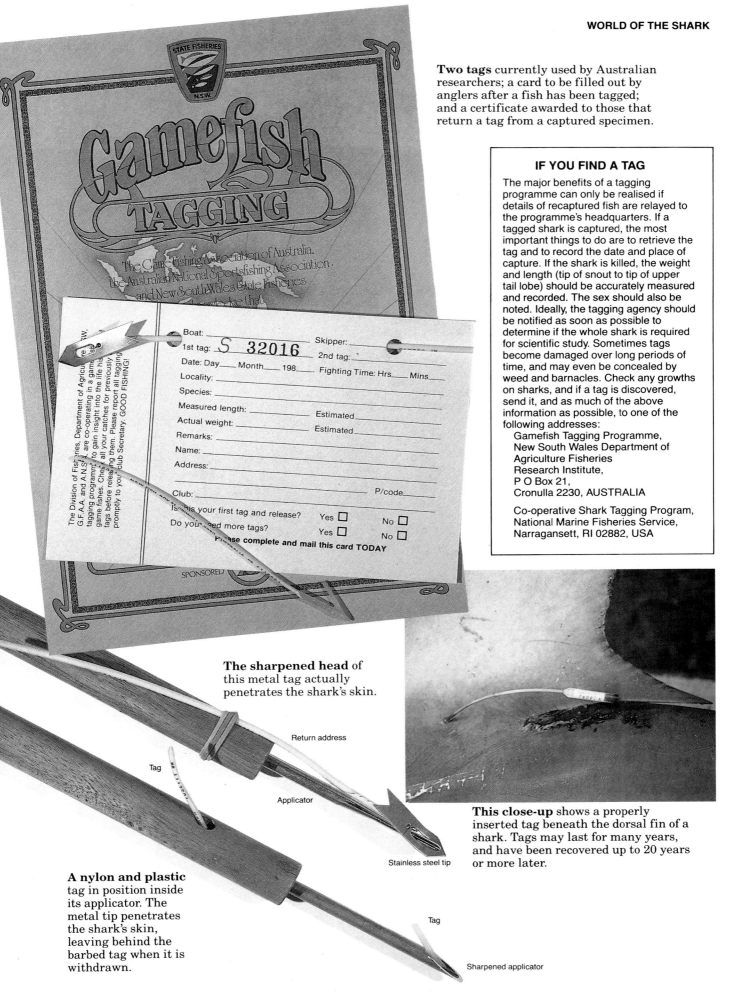

Two tags currently used by Australian researchers; a card to be filled out by anglers after a fish has been tagged; and a certificate awarded to those that return a tag from a captured specimen.

IF YOU FIND A TAG

The major benefits of a tagging programme can only be realised if details of recaptured fish are relayed to the programme's headquarters. If a tagged shark is captured, the most important things to do are to retrieve the tag and to record the date and place of capture. If the shark is killed, the weight and length (tip of snout to tip of upper tail lobe) should be accurately measured and recorded. The sex should also be noted. Ideally, the tagging agency should be notified as soon as possible to determine if the whole shark is required for scientific study. Sometimes tags become damaged over long periods of time, and may even be concealed by weed and barnacles. Check any growths on sharks, and if a tag is discovered, send it, and as much of the above information as possible, to one of the following addresses:

Gamefish Tagging Programme,
New South Wales Department of
Agriculture Fisheries
Research Institute,
P O Box 21,
Cronulla 2230, AUSTRALIA

Co-operative Shark Tagging Program,
National Marine Fisheries Service,
Narragansett, RI 02882, USA

Return address
Tag
Applicator
Stainless steel tip
Tag
Sharpened applicator

The sharpened head of this metal tag actually penetrates the shark's skin.

A nylon and plastic tag in position inside its applicator. The metal tip penetrates the shark's skin, leaving behind the barbed tag when it is withdrawn.

This close-up shows a properly inserted tag beneath the dorsal fin of a shark. Tags may last for many years, and have been recovered up to 20 years or more later.

GREAT WHITE DEATH

A curious great white shark closes in to inspect a potential meal. These sharks will repeatedly ram a cage in a search for something edible, their teeth tearing at the flimsy mesh. On more than one occasion badly shaken divers have had to be evacuated by their companions from damaged cages.

Villain of a hundred stories, hated, abused, misunderstood, the great white shark Carcharodon carcharias *rose to prominence in films such as* Jaws *and* Blue Water, White Death. *It is therefore surprising to discover that, for all its fearsome reputation, very little is known scientifically about this huge fish. Few people have had more experience with them in their natural environment than the photographers Ron and Valerie Taylor, and here Valerie gives some of her own observations about these monsters of the deep.*

'In 1965 Ron became the first person to film a great white shark underwater. He did this by hanging unprotected from the platform at the back of a tuna boat. On board, along with shark victims Brian Roger, Rodney Fox and Henri Bource, was the world famous shark fisherman, Alf Dean. During the trip Alf caught five great whites and taught us how to attract them.

'It was the first and last time that Ron was involved with the mass slaughter of sharks. It was also the first and last time that he worked with great whites without the protection of a steel cage.

'We have been filming whites now for more than 20 years, and we still use Alf's original method of attracting them. Now, however, we have to use far more burley (fresh minced tuna, blood and fish oil), and there seem to be fewer great whites to attract. I believe that unless some protection is given to these beasts, they may be doomed to extinction.

'The great white shark, or white pointer, is the largest potentially

Three consecutive frames (below) from a movie film show the astonishing speed with which a great white shark seizes its prey. The photographs are only four-hundredths of a second apart. In order to attack, the shark changes the whole shape of its head so that the mouth moves to the front, and the snout bends up out of the way. After biting, the mouth moves back to its original position. Even a 5-m (16-ft) great white can easily tear a piece of blubber the size of a large dog from the tough hide of a dead whale.

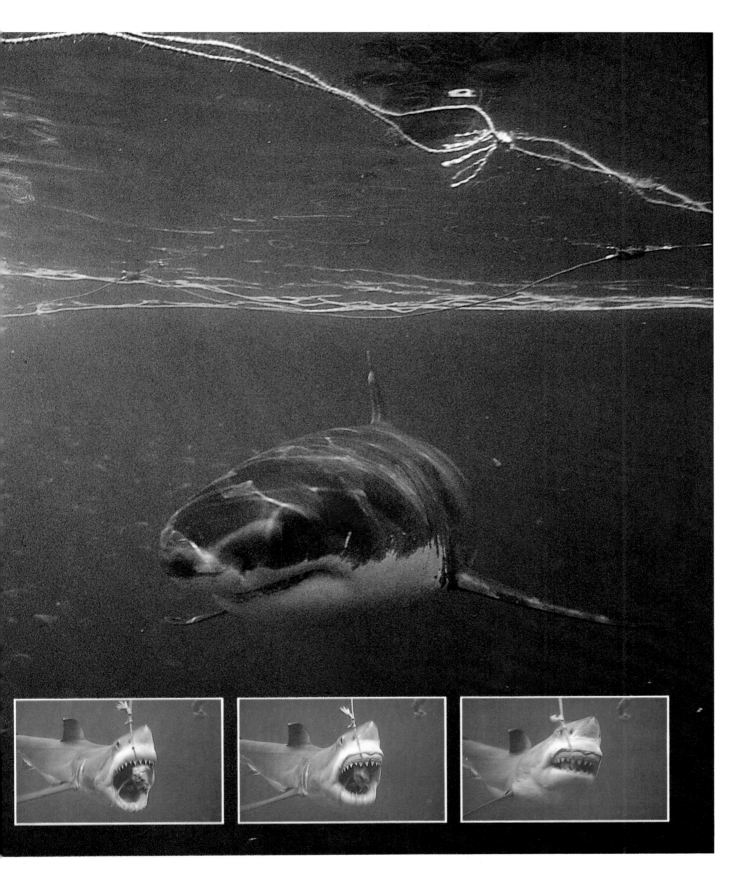

Teeth from the upper and lower jaws of a great white shark have razor sharp serrated edges, specially adapted for slicing through the toughest skin.

dangerous fish in the ocean. It tends to live in temperate waters, although it has been found in almost every ocean on earth.

'We believe that great whites are basically territorial, returning regularly to the same area year after year. The size of their territory may vary according to the richness of the feeding grounds and the number of other large sharks living nearby. On one occasion we tagged a shark which was subsequently caught six months later only 1 km (1.6 miles) away from the tagging site. We have also noticed that we see the same sharks year after year in the same areas. Not all sharks are easily recognisable, but some have unmistakable peculiarities that make them stand out. There seems to be little doubt that sharks remember past encounters with us. When they first come up against our steel cage they test its strength by biting it, trying to get at the diver inside. On a second meeting they hesitate to bite, apparently realising that the steel bars will break their teeth.

'Many sharks, because of their size, need large amounts of food. This is why great whites are often found around seal colonies, rather than near coral reefs.

'Very little scientific research has been carried out on great white sharks. They are not commonly seen unless attracted by baits. Because their diet occasionally includes humans, they can only be studied in their natural environment from a cage. In such circumstances it often seems that the observer, imprisoned behind bars, has become the observed, while the sharks are free to come and go as they wish.

'Most of what we know about great whites is related to their feeding patterns. What they do when they are not looking for food is a matter for speculation. We can only guess as to where they breed, how large they are at birth, how often they eat and how far they travel under normal circumstances.

'Some facts have, however, emerged from our observations. Over the years we have noticed that adult great whites in some areas tend to live in groups according to their size and sex. At Dangerous Reef in South Australia, for example, the average shark we see is a male 3 to 5 m (10 to 16.4 ft) long. Further west, around Streaky Bay, the average shark is a female around 5 to 6 m (16.4 to 19.7 ft) long. Juvenile great white sharks are not seen by divers, yet at certain times of the year they are caught in mesh nets off southern Queensland.

'Along the northern Californian coast, where great whites are very common, the sizes and sexes seem to be more mixed. In this area there are many islands that support a large population of seals.

'A few years ago it was believed that sharks, like most fish, were cold blooded – that their body temperature was the same as that

The monster contemplating this diver is around 3.5 m (11.5 ft) long. While record claims are confused and difficult to verify, it appears that great whites can occasionally reach an astonishing 9.1 m (30 ft) long, and may perhaps grow even larger.

Great whites are the only sharks that lift their heads clear of the water so that they can inspect objects on the surface. They will also emerge partly from the water to secure a good grip on an intended prey.

of the surrounding water. At the request of an American shark expert, Mr Thomas A. Lineaweaver, we always carried two large thermometers with us when diving. Mr Lineaweaver believed that large, fast moving fish and sharks, because of constant muscle action, may generate a body temperature higher than that of the surrounding ocean. He therefore asked us to take the temperature of a living great white shark.

'In January 1973, while we were filming for a television special, a large female great white, 4.3 m (14 ft) long and weighing around 700 kg (1540 lb), became entangled in the steel trace attaching our cage to the boat. This was a perfect opportunity. Once she was tired, we carefully dragged her into shallow water. She struggled half-heartedly, but the steel trace held fast. Ron drove a slender knife deep into her back, then inserted the thermometer. Her body temperature was 22.55°C, and at the time the water temperature was 20.5°C. When we measured the shark's temperature she had been practically motionless for about three hours, so muscle action could hardly account for her higher body temperature (see p.20).

'In 1965, when we first began working with great whites, I thought they were dumb brutes – all brawn and no brains. Now, after working with hundreds of these sharks, I know my original impression was false. Great whites may have small brains, but they seem to use them effectively. They can be taught simple tricks in about 20 minutes – such as to always approach the sunny corner of a boat for food (for filming purposes). Great whites can also see very well, and they are the only sharks to lift their heads clear of the water to inspect surface objects. It is quite unnerving to look up and find a shark watching your every move.'

SHARKS FOLLOW THEIR NOSES

Sharks have an extremely acute sense of smell, and are able to detect very small concentrations of some chemicals – particularly those that are important to them when hunting food. Some experiments carried out in the United States of America have proved sharks to be capable of responding to concentrations of certain chemicals in dilutions as low as one part per thousand million (for more details of this amazing ability see p.32).

Sharks are able to use smell to home in on a potential prey by adopting a very simple strategy: whenever they encounter the smell they simply turn into the current. By constantly swimming and turning upcurrent whenever the smell is detected (diagram right) they are able to find its source very rapidly. Humans and other land animals use very much the same method to find the source of an air-borne smell.

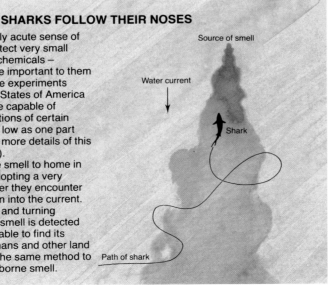

Source of smell

Water current

Shark

Path of shark

THE LARGEST FISH ON EARTH

Whale sharks and their relatives the basking sharks, are the largest fish on earth. However, while basking sharks are common enough to be hunted commercially for their livers, whale sharks, the larger of the two, are rarely seen.

Cavern like, the huge mouth of a 10-m (33-ft) whale shark dwarfs a diver. Fortunately for humans, these giant fish live almost exclusively on microscopic plankton, filtered from the water as they swim slowly along, just beneath the surface of the ocean.

A comparison between a whale shark and a medium-sized passenger car gives some idea of the fish's great size. Whale sharks are about the same size as sperm and humpback whales, but are considerably smaller than blue whales.

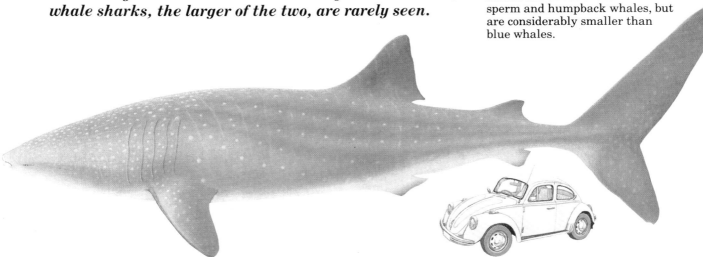

The name often causes confusion in the minds of laymen, but these gigantic fish are definitely sharks. The word whale merely refers to the creature's great size, and they have probably been mistaken for whales on many occasions. Whale sharks may grow to almost 18 m (59 ft) in length, and can weigh up to 40 000 kg (88 200 lb).

Because of their size, it is fortunate that they are plankton feeders. Humans have little to fear from these gentle creatures except for the possibility of having a boat bumped by one as it cruises along just beneath the surface of the sea.

Little is known about the lives of these giant sharks. They have two dorsal fins, a mackerel-shaped tail, small eyes and large gill slits. The body is ridged lengthways in a similar fashion to that of the leopard shark *Triakis semifasciata* and Port Jackson shark *Heterodontus portusjacksoni*. Like these two species, whale sharks also lay eggs, the world's largest. One specimen trawled up from 57 m (187 ft) in the Gulf of Mexico measured 300 x 140 x 80 mm (12 x 5.5 x 3.2 in).

The embryo inside the egg was a well developed spotted replica of the adult. How fast newly hatched sharks grow, or how long they live remains a mystery. Only mature whale sharks have been observed, generally swimming slowly on the surface, their dorsal fins protruding from the water.

Whale sharks are found, although rarely, in most tropical waters. They are seen regularly near the Maldive Islands and the Seychelles in the Indian Ocean. At certain times of the year in the Gulf of Mexico charter boat operators can practically guarantee divers a swim with a whale shark.

The Gulf of Mexico is exceptionally rich in marine life. Plankton clouds the water, attracting all types of filter feeders, including whales and whale sharks. While plankton and small school fish seem to make up the largest part of a whale shark's diet, they have also occasionally been seen eating tuna, when both are sharing schools of bait fish. In each case, the sharks were feeding vertically, moving up and down through masses of small fish, their blunt heads periodically breaking the surface of the water. Tuna that were

feeding on the same fish were seen to be scooped into the sharks' enormous mouths, and although some jumped out, many were apparently eaten.

Whale sharks pay little attention to divers. Probably because of their size, they have little to fear in the ocean.

Since the discovery that they are common in the Gulf of Mexico, riding them has become a popular pastime with the lucky few. The sharks seem to tolerate humans hanging onto their dorsal fins, pectoral fins and tails without complaint. The diver's presence, provided it does not give discomfort to its host, probably causes no more inconvenience than sucker fish do to small sharks.

One group of American divers met a whale shark that actually appeared to like them. She was around 10 m (33 ft) long, and even when they had finished their dive, she followed the charter boat, apparently waiting for the divers to return. This amazing association continued for three days, after which the shark left. It is well known that whale sharks enjoy rubbing against floating objects,

possibly to dislodge unwanted parasites. In this case, there were seven divers riding on the shark, so perhaps she simply enjoyed the constant handling.

Whale sharks are not often seen in Australian waters, but they are sometimes sighted. Diver and photographer Valerie Taylor described one such exciting encounter: 'We had our big thrill one bleak autumn day in the early 1970s. Ron Taylor, myself and a friend, John Harding, were returning from Seal Rocks, two miles [3.2 km] off Sugarloaf Point, New South Wales, when a huge fin broke the surface. At first, because of its size, we thought it might be a great white shark. I remember Ron saying, "No great white can be over 30 ft [9 m] long, it must be something else". We were all staring excitedly at the huge shape ahead of us. In those days, only two other divers had ever ridden and filmed a whale shark. Our delight at this incredible piece of luck made us forget the cold, dismal day and rough seas. This was our chance, and we certainly made the most of it. The shark was a large female, between 35 and 45 ft [10.7 and

THE STRANGE HISTORY OF WHALE SHARKS

This small whale shark was the victim of a collision in the Indian Ocean in 1932.

The first whale shark to be scientifically examined was harpooned in Table Bay, South Africa, in 1828. It was only a small specimen – a mere 4.6 m (15 ft) long – and the skin eventually found its way into a museum in Paris. Since then a number of whale sharks have been rammed by ships or stranded, but only a very few have been properly examined.

Over 20 collisions between whale sharks and ships have been recorded. These are apparently due to the fact that whale sharks are surface feeders, slow moving and seemingly oblivious to danger. Both ships and sharks are damaged in these encounters, the sharks usually coming off worst. It is, however, interesting to note that the skin of a nine-metre (30-ft) whale shark was found to be 102 mm (4 in) thick – the thickest skin found on any living animal.

Various claims have been made for the largest whale shark ever sighted – some reaching an unlikely 20 to 23 m (65 to 75 ft). The largest properly measured specimen, caught near Karachi, Pakistan, in 1949, was 12.65 m (41.5 ft) long.

13.7 m] in length, and we stayed with her for around three hours and travelled roughly 5 miles [8 km].

'She was heading steadily south at about 2 mph [3.7 km/h]. We turned her a few times, but she always swung back onto her original course. We were not using scuba diving equipment so could keep up with her if we swam fast, but our aluminium dingy was a problem. Fortunately the wind was pushing the 14-ft [4.5-m] boat in the right direction, but we still had to constantly pick it up and motor after our shark. Probably the most exciting moment, next to first seeing her, was when John and I rode on her tail. Being dragged through the water at tremendous speed as the tail swept to and fro in a 7-ft [2-m] cycle was a thrilling experience that I shall never forget.

'Eventually cold, failing light, and a strengthening wind drove us from the water. We last saw her, a large dark shape, disappearing through the white caps. Pounding into the worsening weather, we returned to our camp at Seal Rocks, the happiest divers in the world.'

A dwarf shark, barely 100 mm (4 in) long.

THE SMALLEST SHARK

In direct contrast to the whale shark, and shorter at sexual maturity than one of its giant eggs, is the spined pygmy or dwarf shark *Squaliolus laticaudus*. This tiny fish is the smallest known species of shark, reaching a maximum of only 250 mm (9.8 in). It is rarely seen because it lives on the ocean floor during the day, and travels 200 m (660 ft), or more, to feed near the surface at night. Dwarf sharks have numerous luminescent organs on their undersides which makes them more difficult to see from beneath, and therefore less vulnerable to predators.

Tab. vi.

Tænia altera

Meer haub Tænia

Tænia Bellonÿ

Tænia falcata Imperati

Canis Carcharias

Großer Meer hun.

Canis Carchariæ species alia

ein ander art Meer oder Walhunds

Canis Carcharias altus

Part Two

2

Men and Sharks

At a time when
humans have little to
fear from the animal world
on land, sharks still reign
supreme in the oceans. Here is
the fascinating story of
man's relationship with
this most terrible of
all predators.

Illustrations of sharks
from Conrad Gesner's
Fisch-Buch, first
published in 1670.

'A Marveilous Strange Fishe'

Human beings have known of the existence of sharks for as long as they have lived near the sea. To early humans sharks were only one of many dangerous and hostile animals that lived in the world around them. The encroachment of wild animals into the human domain was always alarming; at least sharks kept to the oceans.

Early societies saw the animals and plants around them in symbolic and religious terms. The natural world had no meaning or validity in its own right, but only in its resemblances and analogies to people, and sharks, like many animals, developed a role in various legends. Often they were the basis for the sea-monsters of mythology.

Although human societies knew enough about sharks to be afraid of them, there was no understanding of their natural history. Very little was known of the range of shark species, and there was no sense of classification. Sharks were known by many and changing names.

One of the earliest stories that may have involved a shark is the ancient Greek legend of Andromeda, the daughter of Cepheus, king of Ethiopia. Andromeda's mother boasted that her daughter was more beautiful than the Aegean nymphs known as Nereids. Poseidon, god of the sea, was angered by this and sent a sea-monster to ravage the coastline. The monster would only be placated by the sacrifice to it of Andromeda herself. Andromeda was then chained to a rock but was rescued by Perseus who slew the monster and married Andromeda.

The Biblical story of Jonah and the whale – if it can be taken literally at all – is now thought to refer to a shark rather than a whale. The original Hebrew version of the story uses the word *tanninum* which means any large sea animal. Traditionally this was thought to have referred to a whale because Saint Matthew, in the New Testament, mentions the story of 'Jonah and the whale'. Both whales and sharks were seen in the Mediterranean at that time.

There are a number of reasons why a shark, especially a great white shark, was more likely to have swallowed Jonah than a whale. The white shark has a less rigid diet than the largest toothed whales, such as the sperm, and is a more opportunistic feeder. They can also swallow large animals whole. One white shark was found to have an entire 50-kg (110-lb) sea lion in its stomach. The sixteenth century French naturalist Guillaume Rondelet reported finding a whole (and fully clad) sailor in the stomach of a great white. He believed that this shark was the 'fish' in the story of Jonah.

Some sharks also have a curious digestive system which allows them to store food in their stomachs for long periods without breaking it down. One tiger shark kept two intact dolphins in its stomach for over a month. Sharks can also regurgitate the entire contents of their stomachs at will.

The first historical reference to sharks is by the Greek writer Herodotus, known as 'the father of history'. His *Histories,* written in the fifth century BC, was the first great prose work of European literature, and described the struggle of the Greek city states against the Persian empire. In 492 BC, the Persian fleet was hit by a violent gale and Herodotus recorded that: 'a great many of the ships were driven ashore and wrecked on Athos [in northeastern Greece]. Indeed, the report says that

The 'great fish' that swallowed Jonah was more likely to have been a shark than a whale. Great white sharks are quite capable of swallowing a man whole, and there is even a report from 1758 of a sailor being swallowed by a shark which was forced to vomit him up, unharmed, shortly afterwards.

A mosaic from the Roman town of Pompeii, dating from about the first century AD, illustrates some of the life to be found in the Mediterranean Sea at that time. Included are two sharks (bottom left and top right).

something like 300 were lost with over 20 000 men. The sea in the neighbourhood of Athos is full of monsters so that those of the ships' companies who were not dashed to pieces on the rocks were seized and devoured.'

The first attempt to understand sharks as animals rather than mythologically was made by the philosopher and scientist Aristotle, who lived in Greece a century after Herodotus. His most important work on biology was the *Historia Animalium (Enquiry into Animals)* written about 330 BC.

Aristotle lived for many years by the sea in Macedonia and gathered a great deal of information about fishes and their habits from local fishermen. His knowledge of the fishes of the Mediterranean was greater than that of any other ancient writer and greater, in fact, than any naturalist until the sixteenth century. He was the first to describe the distinction between the bony and cartilaginous fishes. Although his fishes are described separately without any attempt at classification, Aristotle gathered the sharks and rays into a group he called *Selache,* a name still used in various forms by modern ichthyologists. He noted that sharks, unlike other fishes, have uncovered gills, and that, with some exceptions, they bear live young.

Aristotle was also the first to record that sharks mate by internal fertilisation (although he also wrongly assumed that other fish do so as well): 'All fishes with exception of the flat selachians [skates and rays] lie down side by side and copulate belly to belly ... Again in cartilaginous fishes the male, in some species, differs from the female in the fact that he is furnished with two appendages hanging down from about the exit of the residuum, and that the female is not so furnished.'

Although Aristotle was right

more often than he was wrong, he also started a myth about sharks that persists to this day. He noted that 'the mouths of sharks are placed not at the front but on the underside of the head ... so that these fishes turn on their back in order to take their food. The purpose of Nature in this was apparently not merely to provide a means of salvation for other animals, by allowing them opportunity of escape during the time lost in the act of turning – for all the fishes with this kind of mouth prey on living animals – but also to prevent these fishes from giving way too much to their gluttonous ravening after food ... an additional reason is that the projecting extremity of the head in these fishes is round and small, and therefore cannot admit of a wide opening.' Aristotle was wrong. Shark jaws are, of course, hinged to allow a near vertical gape and

A 4th century bust of Aristotle, whose observations of sharks and their behaviour, compiled in about 330 BC for his *Historia Animalium*, remained the most thorough study of the subject for almost 2000 years. Aristotle was also responsible for the myth that sharks must turn on their backs to attack.

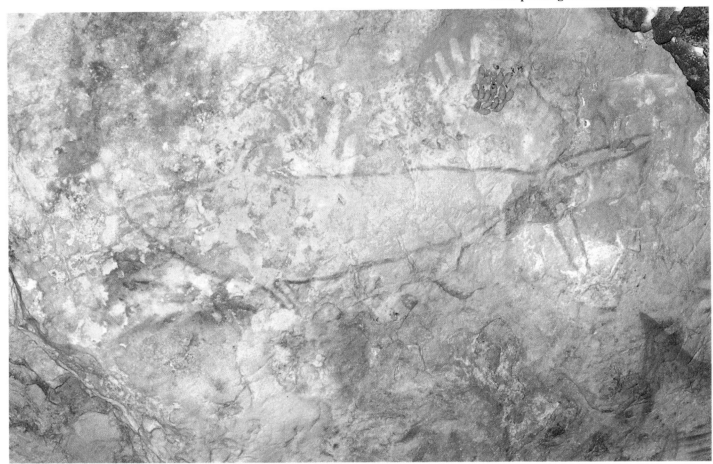

sharks can and do bite right side up.

Over 400 years after Aristotle's *Historia Animalium,* another popular work on the natural world was compiled by Pliny the Elder, a Roman administrator and writer. In his encyclopaedic *Natural History,* written in about 77 AD, he described in colourful terms the dangers of dog-fishes, a general term for smaller sharks. 'Divers have fierce fights with the dog-fish; these attack their loins and heels and all the white parts of the body. The one safety lies in going for them and frightening them by taking the offensive; for a dog-fish is as much afraid of a man, as a man is of it.'

Other classical authors provided misinformation about sharks which persisted in some cases into the nineteenth century. Oppianus, a third century Greek poet from Cilicia, wrote a long fishing epic called *Halieutica*. In it he developed a myth about blue sharks as ideal parents which is entirely inaccurate.

> *'Thus the Blue sharks, secure*
> *from chasing foes,*
> *Within their widen'd mouths*
> *their young enclose.*
> *Beneath the circling arch they*
> *fearless hide,*
> *The bulky forms drive on the*
> *rising tide.*
> *Of all oviparous kinds that*
> *throng the seas,*
> *The fond Blue sharks in tender*
> *care surpass.'*

Sharks are, of course, found in all the world's oceans and other societies and cultures dealt with them in their own particular myths and practices.

The pearl divers of Sri Lanka,

to this day, place great reliance on shark-charmers, a custom that was first noticed by the Venetian traveller Marco Polo in 1298. 'There are moreover in that bay of the sea [the Gulf of Mannar] a multitude of great fishes,' he wrote, 'which would kill the fishers going down to the sea. But provision against that danger is made by the merchants in this way. The merchants take certain magicians with them who are called braaman, who with their enchantments and diabolical art control and stupefy those fishes so that they can hurt no one. And because this fishing is done by day and not by night, those magicians make spells by day which they break for the following night; for they are afraid lest anyone go down by stealth without the leave of the merchants into the sea by night to take pearls.'

The power of the shark charmers was supposed to be hereditary. 'At one time the charmer used to be allowed a percentage of one oyster per day from each diver but this has been commuted to a

Three mythical sea creatures from the late fifteenth century – a sea-cow, a sea-dog and a sea-horse. Small sharks were given the name sea-dogs or dogfish simply because they have sharp teeth, and popular imagination made them the nautical equivalents of dogs.

money payment', reported an article in the *Cornhill Magazine* in 1866. 'Accidents have never been known to occur on the pearl divers from sharks, which is of course attributed by the superstitious natives to the wise charming of the charmer.'

In the Pacific Ocean, shark

fishing has always been important to the economies of the island communities, and as a result has been overlaid with religious symbols and ceremonies. In contrast to the European idea of sharks as evil monsters, Pacific islanders have had a more ambivalent attitude; in many parts of Polynesia, for example, shark worship is still common.

Hunting the *kapeta* or dog-shark by New Zealand Maoris was a ritual controlled by priests from the shore. Up to 1000 men would set out to hunt the sharks on two specified days of the year. Other shark species could be caught at any time. The Society Islanders, in French Polynesia, also worshipped one shark, the blue, at the expense of others. 'Temples to these sharks were erected,' reported the missionary William Ellis in the late nineteenth century, 'and offerings were presented to the deified sharks while fishermen and others, who were much at sea, sought their favour.'

There are a number of legends of humans reincarnated as sharks, and the Hawaiians tell the story of the *mano-kanaka* or shark men – sharks who take human form to create mischief on land. In Hawaii and on the Solomon Islands there is evidence that people were once sacrificed to sharks. In Tonga, shark hunters would go to great lengths to obtain occult support. They would be celibate the night before going to sea, and while they were away the whole village would have to remain totally quiet.

Although the arrival of

An Australian Aboriginal bark painting from north-east Arnhem Land shows a fresh water shark, probably one of the whalers (family Carcharhinidae). Sharks were important to the Aborigines – as they were to many native people around the world – as a source of food, and because of the valuable oil in their livers – which is shown in this painting.

For medieval Europeans, who were only just starting to explore the world, the oceans teemed with hostile monsters. Some of the strange creatures in this illustration, which dates from about 1500, were clearly based on hazy descriptions of sharks (such as those labelled K, S and N).

Christianity in the nineteenth century reduced shark worship in the Pacific, many features of it remain in the more remote islands today. In the 1930s, Hawaiian shipwrights would try to propitiate sharks during a ship's maiden voyage. This custom was common in east Africa as well, where boat-builders still anoint the wood of new vessels with hammerhead shark oil to bring a good wind and a successful voyage.

Sharks also occur in Japanese legend. One of the traditional Japanese deities is a storm god known as Shark-Man. In one legend Shark-Man was rescued by Totaro whom he then made very rich. Totaro, however, wanted more – to marry a girl named Tamara. He asked Shark-Man for jewels for his bride and Shark-Man, after reproaching him for his greed, wept blood into a dish. The tears crystallised into ten thousand rubies. Another Japanese legend involved a warrior who performed amazing feats of strength, one of which was to swim through the ocean with a shark under each arm.

In India, the god Vishnu is sometimes represented as coming out of the mouth of a monstrous fish, probably a shark. Vishnu is revered for having saved the *Vedas,* the sacred religious texts of India, when the entire world was drowned – an obvious parallel with the legend of Noah.

Many American Indian tribes have given sharks a place in their folklore. Some tribes, reported the herpetologist Lawrence Klauber, referred to rattlesnakes as 'the little sharks of the woods'. The Tlingit Indians used shark crests as the curved emblems for tribal clans. In Central and South America shark images have been found on ancient Indian pottery, and figurines depicting shark attack scenes were also made.

After the close of the classical age in Europe, curiosity about geography and natural history waned, and few observations of sharks were recorded. Europe closed in on itself and took little notice of the world outside. Marco Polo mentioned sharks in the seas around Ceylon in the account of his journey to China in the thirteenth century, but it was not until the great voyages of exploration beginning in the fifteenth century that European chroniclers took an interest in sharks once again.

On Christopher Columbus's voyage to the West Indies in 1502, his ships were at one stage surrounded by sharks. 'It was frightening, especially to those who believe in omens,' recorded his son Ferdinand in a biography of his father. 'For just as some say that vultures recognise the presence of a corpse by its smell many leagues away, so some believe that sharks have the same divinatory power.'

Ferdinand Magellan noted the presence of sharks on his voyage around the world between 1519 and 1522. Off the west coast of Africa during a calm he recorded that 'great fish called tiburoni [sharks] approached the ship. They have terrible teeth and eat men when they find them alive or dead in the sea. And the said fish are caught with a hook of iron ... but they are not good to eat when large.'

Roger Barlow, an English geographer, agreed with Magellan in 1540 about the quality of shark flesh – 'the meete is very tough and unsaverie', he reported. However, Peter Martyn Anglerius, a Dutch navigator in the West Indies wrote in the 1550s that 'the leaste of these fyshes [sharks] are most holsome and tender. When they have slayne this fyshe they [sailors] cutte the body therof in smaule pieces and put it to drye, hangynge it three or foure dayes at the cordes of the sayle clothes.' Anglerius also noted how sharks 'coome furth of the sea and enter into the ryvers ... they are very dangerous in certeyne washynge places or pooles by the ryvers sydes.'

Many unreliable legends still surrounded the European view of sharks. Olaus Magnus, a noted Swedish historian, wrote in 1555 of the malevolent 'sea-Dogfish that will set upon a man swimming in Salt-Waters so greedily ... that he will sink the man to the bottom, not only by his biting, but also by his weight; and he will eat his more tender parts, as his nostrils and fingers.' Magnus went on to describe how the ray (a cousin of the shark) will come to the rescue of the

The illustrator working on Sir Thomas Herbert's account of his voyages, published in 1638, managed to get two of the hazards facing unfortunate sailors into the same engraving – sharks and tornadoes.

FAITHFUL COPIES

By the end of the eighteenth century enough information had been gathered by European explorers to enable scholars to assemble the first comprehensive guides to the plants and animals of the world. The first major work on fish – which included the 15 species of sharks then known – was published in Germany between 1785 and 1795, under the title *A Complete Natural History of Fish*. Its author was Marcus Eliezer Bloch.

Bloch was born in Ansbach, near Nuremburg, Germany in 1723. He was the son of poor Jewish parents and received hardly any education as a child. At the age of 19 he could still not read, but despite this managed to become tutor to the household of a barber in Hamburg. While working there Bloch realized that he needed to educate himself and, with the help of relatives, he went to Berlin to study medicine. He eventually gained his degree in Frankfurt, and later returned to Berlin.

Bloch was fascinated by fish all his life, and eventually came to be regarded as the most important ichthyologist of the 18th century. Both of his major works – a four-part monograph on the fishes of Germany (1781) and the nine-part *Natural History* – were published at his own expense.

The *Natural History* contained 216 colour plates which became a standard source of reference for artists and publishers for many years afterwards. Even errors, such as the jaws of the great white shark which were shown upside down in Bloch's book, were perpetuated in dozens of other publications for many years afterwards.

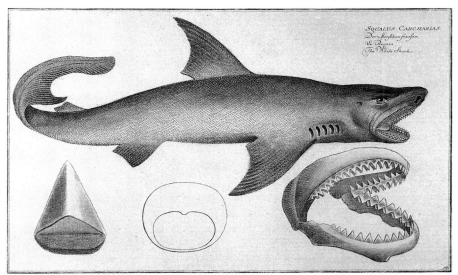

The great white shark from Marcus Bloch's Complete Natural History of Fish *(1785-95).*

Bloch's shark *transferred to a South Seas island setting.*

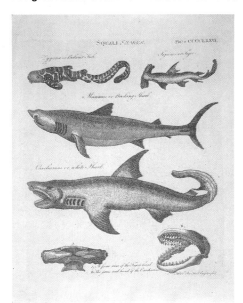

The white shark's *inverted jaws were faithfully reproduced in this copy.*

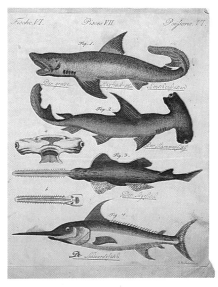

Any comparison *with a real shark would have revealed serious errors in the fin shapes on the white shark.*

man and drive the shark away – another myth that survived into this century.

Until the mid-sixteenth century the word shark in English referred only to human low life. Small sharks were known as dogfish and large ones as *tiburon* – the Spanish word for shark. The current meaning of shark did not come into the language until 1569, after a specimen captured by the explorer and slave trader Sir John Hawkins was displayed in London. 'There is no proper name for it (a marveilous strange Fishe) that I knowe,' said one observer, 'but that sertayne men of Captayne Hawkinses doth call it a sharke.' Other words such as tiburon, haye, requiem and sea-dog continued to be widely used for many years.

Because natural historians had not advanced much further in the classification of sharks since Aristotle's time, new shark-like animals were discovered and described without any attempt at understanding their relationships

Curious passers-by examine the carcase of a great white shark at a seaport somewhere in France. Great whites are found in most of the world's oceans, and in the Mediterranean Sea.

with known fishes. They were always seen in human terms – normally unpleasant ones.

The French naturalist Guillaume Rondelet wrote in 1558 of a 'sea-monster' taken in Norway after a great storm which he named monk-fish since 'it had a man's face, rude and ungracious, the head smooth and shorn. On the shoulders, like the cloak of a monk, were two long fins in place of arms and the end of the body was finished by a long tail.' The fabulous creature described by Rondelet is today known as the angel shark.

The earliest reliably-documented shark attack occurred in 1580 somewhere between Portugal and India. A sailor fell overboard from his ship and an eyewitness described, as the sailor was hauled by a rope towards the ship, how 'there appeared from below the surface of the sea a large monster called Tiburon; it rushed at the man and tore him to pieces before our very eyes. That was a very grievous death.'

By 1600 the word shark had spread into fairly common English usage. Shakespeare used it in the recipe for witch's stew in *Macbeth* (written in 1606):

> *'Scale of dragon, tooth of wolf*
> *Witches' mummy, maw and gulf*
> *of the ravin'd salt-sea shark.'*

Sharks and shark-related customs around the world were described by a growing and far-flung band of European travellers. Samuel Purchas in 1617 noted that along the Ganges delta in India there were 'many Fishes called Sea-Dogs. They which are weary of this world, and desire to have a quick passage to Paradise, cast in themselves here to be devoured of these Fishes.'

Twenty years later Sebastian Manrique described the customs of

---

THE FRUITS OF EXPLORATION

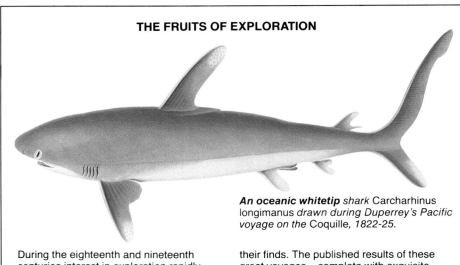

An oceanic whitetip shark Carcharhinus longimanus *drawn during Duperrey's Pacific voyage on the* Coquille, 1822-25.

During the eighteenth and nineteenth centuries interest in exploration rapidly turned into a fever. Ships from many seafaring nations – including France, Britain, Russia, Germany and the United States of America – explored, charted, mapped, claimed and colonised land in the furthest corners of the earth.

Most expeditions carried with them trained scientists and artists to document

their finds. The published results of these great voyages – complete with exquisite illustrations of the plants and animals that they found – gave people in the western world their first accurate view of the creatures around which so many myths had once been woven.

Mermaids turned into dugongs, krakens into cuttle-fish and sea monsters into whales and sharks.

Hindu pilgrims at the mouth of the Hooghly River. 'As soon as they have made this solemn vow [to the priests] they enter the sea up to their breasts and are very soon seized and devoured by certain ferocious male and female sea-monsters which we call sharks.'

Gradually sharks ceased to be only fearsome monsters, and a growing curiosity appears in the literature. 'I was told in Batavia and elsewhere,' wrote Francois Leguat in 1691, describing a number of shark remedies, 'that the Brain of a Shark had a Virtue in it, which made Womens' Pains in Child-Bed not so racking to them as they generally are.' In 1730 Monsieur Pomet, Louis XIV's chief druggist, asserted that 'there is found in the Shark's head two or three spoonfuls of brain that is as white as snow, which being dried, reduced to powder, and taken in white wine, is excellent for the gravel [kidney stones]'.

Later Pomet went on to sum up the popular view of sharks – a view that has not changed much to this day: 'He is one of the most gluttonous animals in the world; nothing comes amiss to him; tho' it be a log of wood he will swallow it, provided it be but greasy, for he swallows without chewing. He is furious and bold, and will throw himself to the shore, and remain almost on dry land, that he may have the opportunity of catching the passengers. Sometimes he will bite at the very oars with his sharp teeth, for rage and madness that he cannot get at the men which are in the boat.'

From the eighteenth century onwards accounts of sharks start to show a gradual departure from the assumptions of the past. Naturalists were beginning to see animals as a field detached from the human, to be properly studied and documented in their own right.

In 1771 Peter Orbeck on a

Sailors feared and hated sharks with good reason. Sharks regularly followed ships in the hope of eating any garbage that was thrown overboard, and their presence must have been constantly at the back of every man's mind. Persistent myths reinforced the fear: 'But a far worse character attaches itself to the shark,' wrote one observer, 'which is his preference for human flesh: of all other food it is this that he prizes ...' It is therefore scarcely surprising that shark fishing was a popular sport whenever there was little else to do on board. Even Napoleon Bonaparte (above) was treated to the spectacle of a captured shark while aboard the *Bellerophon* after his surrender in 1815.

Two of the *Challenger's* dogs look on in amazement as a shark is hauled aboard. Everything encountered during the expedition was subjected to intense scrutiny by scientists.

Naturalists aboard the *Challenger* (see box below) were lucky enough to obtain three specimens of the rare frilled shark *Chlamydoselachus anguineus* while they were in Japan. The specimens – two males and one female – were carefully dissected and the structure of their internal organs recorded in painstaking detail in drawings published later in one of the expedition's 50 volumes of reports.

The heart of a frilled shark with its right and left sides marked (r and l).

A section through the spiral valve, which contained 35 gyrations and was 16.5 cm (6.5 in) long.

The structure of the spiral valve (intestines) of the shark in cross-section.

voyage to China reflected on the problems of studying sharks. 'We caught the dogfish today, which is reckoned the most voracious animal of prey. Authors have already described several kinds of them, though not very clearly. The reason thereof is probably that some sorts are nowhere to be found but in great seas, where they can be but seldom examined by inquisitive people: whence all sorts are called by the same name because they look alike at a distance. Very seldom does the opportunity offer of comparing several sorts together, that specific marks might be ascertained, which otherwise is difficult, as their fins do not constitute the only difference.'

Today there are thought to be about 344 species of sharks. At the end of the eighteenth century most of these were still unknown. The German ichthyologist Marcus Bloch (see box p 67) reported in 1795 that there were 14 or 15 known shark species. The largest shark of all, the giant but harmless whale shark, was not discovered until 1828.

Slowly and painfully a body of information about sharks was built up. The 1844 edition of Oliver Goldsmith's *History of the Earth and Animated Nature* (originally published in 1774) stated that there were 30 different species of sharks.

A GREAT VOYAGE OF SCIENTIFIC DISCOVERY

By the second half of the nineteenth century most of the world had been mapped, and scientists had embarked on the vast job of systematically examining and classifying objects in the natural world. However, while much was known about plants and animals that lived on the land, almost nothing was known about those in the oceans. In about 1850 it was even possible for Edward Forbes, Professor of Natural History at Edinburgh, to state categorically that no life at all existed below 548 m (1800 ft).

All this changed in 1872 when HMS *Challenger* set sail from Portsmouth in England on a voyage that was to found the science of oceanography. For almost four years the *Challenger* cruised the oceans of the earth covering an astonishing 127 634 km (68 890 nautical miles). Many thousands of new species of marine plants and animals, including sharks, were discovered and carefully described.

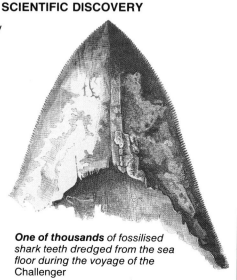

One of thousands *of fossilised shark teeth dredged from the sea floor during the voyage of the* Challenger

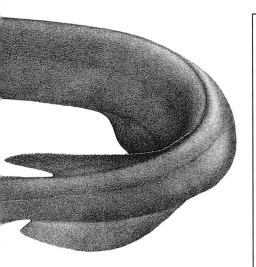

An external view of the claspers belonging to a male frilled shark.

The skeleton of a clasper, showing the cartilaginous rod inside it.

THE SHARK IN FICTION

The popular view *of sharks, as cunning and malevolent, is expressed in this British comic.*

Jaws still menaces visitors to the back lots of Universal Studios in Hollywood.

Sharks have not featured greatly in western art and literature. Representing, as they do, the grisly and ugly, with no redeeming 'human' or interesting characteristics, they have been reduced to symbols of terror in comic books and popular fiction.

Herman Melville deals in passing with sharks in *Moby Dick*. 'De god what made de shark must be one dam injun', says Queequeg, the harpoonist. Mark Twain in *Following the Equator* tells the fanciful tale of how Cecil Rhodes founded his fortune after recovering a copy of the London *Times* from the stomach of a shark caught in Sydney Harbour. News in the paper of a rise in the price of wool, received 40 days ahead of the fastest steamer, allowed Rhodes to make a financial killing. In *The Old Man and the Sea* Ernest Hemingway describes a mako, one of the sharks that destroys the old man's great fish.

Sharks appear in the writings of Melville, Twain and Hemingway as secondary figures, but in 1974 a novel appeared in which the main protagonist was a great white shark. Peter Benchley's book *Jaws*, and the film that followed, released a tidal wave of public concern about the 'shark menace', on a scale that had not been seen since the New Jersey attacks of 1916 (see p 120). Dozens of books and magazines on the subject appeared almost overnight. A second novel and two more films followed. For humans who have tamed and subjugated so much of the natural world, sharks obviously remain as one of the last of the fearsome unknowns – a peg on which to hang unconscious fears.

The *History* gives details of the geographic range and feeding habits of each shark and throws doubt on some persistent myths, such as the story of the blue shark protecting its young in its mouth. However it gives credence to other enduring myths: 'Along the coasts of Africa where these animals [sharks] are found in great abundance,' Goldsmith stated, 'numbers of the negroes who are obliged to frequent the waters are seized and devoured by them each year. The people of these coasts are firmly of the opinion that the shark loves the black man's flesh in preference to the white, and that when men of different colours are in the water together it always makes choice of the former.' Other writers also asserted that sharks preferred Asian rather than European flesh. Shark researchers today give little weight to theories that sharks have a preference for any skin colour among humans.

Until the nineteenth century it was accepted that all animals had been created some time at the beginning of the world, and had remained unchanged in form since then. But even before Charles Darwin published his *Origin of Species* in 1859 the idea that animals changed over time, and that some species became extinct, was a widely discussed theory among naturalists. Today it is known that sharks come from an ancient line of fishes and, due largely to their success within their ecological niche, have changed only slowly, if at all, over the last 300 million years.

Large areas of the natural history of sharks, however, remain a mystery even today. Many people are still obsessed by them – they are one of the few remaining creatures on earth that are outside the control of humans. Because creatures in the oceans are harder to study than those on land; because sharks are often hostile and dangerous; and because most sharks do not breed or behave naturally in captivity, many aspects of shark behaviour will remain a fascinating enigma for a long time to come.

A shark's eye view of the bathers at a beach. Although sharks generally prefer deeper water, most attacks at bathing beaches take place in water less than 2 m (79 in) deep, simply because that is where most people are found.

THE RISK OF ATTACK

Just what are the chances of being attacked by a shark? Some curious facts emerge from an attempt to answer this question. One of the most striking is the shark's apparently overwhelming preference for males – an astonishing thirteen males are bitten for every one female attacked.

The most authoritative analysis of shark attacks is the 1973 report compiled by Dr H. David Baldridge entitled *Shark Attacks Against Man*. In this report Baldridge makes a computer analysis of 1165 reported attacks, considering every possible aspect of shark attack. The 1165 cases make up what is known as the International Shark Attack File. This file was compiled as a result of a United States Navy sponsored meeting of 34 civilian scientists from all over the world. The meeting took place in New Orleans in April 1958 and its original purpose was to seek ways to develop better shark repellants.

Early in the meeting it became apparent that what was known about sharks and their behaviour was woefully inadequate. To help remedy this a group of six scientists was formed into what was called the Shark Panel, with Dr Perry W. Gilbert as chairman. The panel was affiliated with the American

Institute of Biological Sciences and funded by the United States Navy.

The panel was to act as a clearing house for information and to promote basic research on sharks. With financial support from the United States Navy the International Shark File was started with panel member Dr Leonard Schultz of the Smithsonian Institute in charge. In the ten years of its existence (1958-67) Dr Schultz and his workers amassed records of 1652 attacks, some with much detailed information, but many, unfortunately, were only brief newspaper reports. This work ended in 1967 because Navy funding was stopped. Navy funding was available for analysis of data in the files, however, and in 1968 David Baldridge, a United States Navy officer, took the files to the Mote Marine Laboratory in Sarasota, Florida, where he began the task of analysing the data.

The attacks in the file were not

Lone swimmers in deep water stand the greatest chance of being attacked. The fact that more males than females venture into deep water near beaches explains the fact that the ratio of males to females attacked at distances greater than 65 m (213 ft) from the shore is an astonishing 31 to 1.

confined to the years between 1958 and 1967. Records from newspapers and other sources going far back in history were also included. The earliest attack reported was that on a sailor who fell overboard in 1580 while travelling from Portugal to India. Few details of this attack are known, not even the sailor's name, only that he was torn to pieces by a large tiburon - the Spanish and Portuguese word for a shark. Most of the attacks in the file, about two-thirds, happened after 1940, although none were reported from World War II. The reason for this is that Allied medical records list shark attacks as 'unspecified animal bites' and there is no possible way data on shark attacks can be identified. In 487 of the recorded attacks the information is so sparse and unreliable that serious doubt is raised as to whether an attack actually happened, and some hoaxes were found. Thus, only 1165 were analysed in the computer. In spite of this, much valuable information has been obtained from the file which enables the following basic questions about attacks to be answered in some detail.

Which is more likely to be attacked, a male or a female? Not too surprisingly the answer is male. What is surprising is the

Seals are a regular source of food for several shark species, including the great white and the tiger. Their resemblance to wet-suited divers is quite striking.

Surfboard riders, especially when silhouetted against the sky, could easily be mistaken for a seal or a turtle – both of which are eaten by some sharks.

A CASE OF MISTAKEN IDENTITY?

Several species of sharks – notably the great white, the tiger and the mako – are all known to feed on seals and sea lions. In some areas — such as off the southern coasts of Australia and northern California – seals are a major component in the diet of some sharks. Tiger sharks have also been known to eat large sea turtles, either whole or in pieces. The tough shell provides little protection against the sharp teeth and powerful jaws of this species.

Several researchers have pointed out a striking similarity between the shape of a wet-suited diver, complete with flippers, and that of a seal. They have suggested that attacks on divers, particularly in murky water, could simply be the result of a mistake on the part of the shark. The fact that a shark will often lose interest in a human victim after the initial bite, despite the presence of blood, seems to support the theory.

In one Australian attack, the diver Henri Bource, clad in a wetsuit, was swimming with other people among a group of seals when he had his leg torn off. This attack was blamed on either a great white or a tiger shark estimated to be between 3.7 and 4.3 m (12 and 14 ft) long.

This theory has also been extended to explain attacks on surfboard riders. Viewed from beneath, a short surfboard, with its rider's flippers projecting from the back, and his head and arms from the front, could easily be mistaken for a seal or turtle. The noise of paddling, and also that made by the board's fins as they slice through the water at speed, may all help to attract the attention of passing sharks.

FACTORS THAT HELP TO DETERMINE THE RISK OF ATTACK

The seas that wash these beautiful, deserted beaches on the coast of southern Australia are home to many species of sharks, including the tiger and great white. While bathers here are no more at risk than they would be anywhere else in tropical or temperate waters around the world, the possibility of an attack still exists. The labels below summarise some of the findings that emerged from Dr David Baldridge's analysis of shark attack records which were gathered in the United States of America during the 1950s and 60s for the Shark Attack File.

Dr Baldridge reached a number of conclusions about the possible reasons for shark attacks, and these have been expressed in some simple precautions that can be taken by those who consider themselves to be a risk (see box page 81). In essence, the findings may be distilled into the following simple conclusions: Sharks are efficient predators and people who enter their realm run a risk (albeit slight) of being bitten. However, potential victims may draw comfort from the fact that sharks do not seem to be particularly partial to people as food.

Divers seem to be particularly vulnerable to shark attack. Twenty-five per cent of the victims identified in the Shark Attack File were involved in some underwater activity, and of those, 43 per cent were free-diving, with or without a snorkel. The fact that diving takes people into deep water, and that enthusiasts are often handling bleeding and injured fish, obviously makes them attractive to sharks.

Water depth is closely linked to the frequency of shark attacks, but only because most people are found in shallow areas. Sixty-two per cent of incidents in the Shark Attack File (in which water depth was recorded) took place in less than 1.5 m (5 ft) of water. Of the remaining attacks, 12 per cent took place in water from 1.8 to 3 m (6 to 10 ft) deep; 11 per cent from 3.3 to 6 m (11 to 20 ft); 5 per cent from 6.4 to 9 m (21 to 30 ft); 3 per cent from 9.4 to 12 m (31 to 40 ft); 6 per cent from 12.5 to 46 m (31 to 150 ft) with the remaining one per cent taking place in water more than 46 m (150 ft) deep.

Sea conditions probably have more influence on whether people go into the water, rather than on the behaviour of sharks. The Shark Attack File showed that 69 per cent of attacks took place in calm water; 19 per cent in surf; 7 per cent in choppy water; 3 per cent in a swell and only 2 per cent in severe sea conditions.

Human females do not seem to be as attractive to sharks as males, although the reason for this is not at all clear. The ratio of males to females attacked is 13.5:1. Observations have shown that males are generally more active than females in the water, and this may be a factor in explaining the difference. It is also a fact that males tend to venture into deeper water than females.

What the victims were doing at the time of an attack did not seem to be very significant. In 34 per cent of cases they were swimming; in 23 per cent diving; in 20 per cent wading and in 8 per cent surfing (either with or without a board). Splashing, playing around or unusually loud noises were only recorded in 14 per cent of cases. These percentages have been found to agree roughly with the proportions of people involved in these activities on any normal day at the beach.

Water clarity seems to have little influence on the incidence of shark attacks. An analysis of cases in the Shark Attack File, where water clarity was mentioned, showed that roughly half of the attacks took place in clear water and half in cloudy water.

Nearshore waters were the places where almost half of the recorded attacks took place. In cases where a note was made of position of the attack site in relation to breaking waves, the numbers were equally divided between those that took place just beyond the surf line and those that took place inside the surf line.

Distance from the shore seems to play a part in determining whether a bather is at risk or not. The figures show that there is an increase in the proportion of bathers attacked the further one goes from the shore. Over half (51 per cent) of the recorded attacks took place less than 60 m (200 ft) from the water's edge, but this is to be expected because it is in this area that most bathers are found. Twenty-seven per cent of the remaining attacks took place between 60 and 1600 m (200 ft and one mile) from the shore – in a zone where there must be considerably less than half the number of bathers found closer in. The remaining 22 per cent occurred in open sea. There is no minimum distance for attacks – sharks have been known to beach themselves.

The background photograph is of the western bays of Wilsons Promontory, on the southern coast of Australia. In the foreground is Oberon Bay, and in the far distance the shore curves around to Cape Liptrap.

Most bathers at a beach stay close to the shore. Studies of people at beaches have shown that 90 per cent remain in water that is between ankle and neck deep, and that 81 per cent do not venture more than 30 m (100 ft) from the shore.

Inlets *are often used by sharks as breeding grounds, and they are therefore potential attack sites – a claim that is borne out by the records. Eighteen per cent of recorded attacks took place in or near harbours, docks, wharfs, jetties, bays, rivers and river mouths. The fact that garbage is often found in these areas may also help to increase the risk.*

Channels *and areas where the water deepens suddenly are best avoided by swimmers. All such places allow sharks to stalk bathers without being seen themselves.*

0 – 15 m (0 – 50 ft) Thirty-one percent of attacks took place in this zone.

15 – 30 m (51 – 100 ft) Eleven per cent of attacks took place in this zone.

30 – 45 m (100 – 150 ft) Six per cent of attacks took place in this zone.

45 – 60 m (150 – 200 ft) Three per cent of attacks took place in this zone.

60

45

30

15

0
metres

75

While there is no conclusive experimental evidence that sharks respond to particular colours, some experienced divers are sure that they do. Yellow is often named as one colour that will attract great white sharks, and is worn here for that purpose.

Divers carrying injured or bleeding fish are certain to be of great interest to hungry sharks (right). Although sharks are generally discriminating about what they bite, there is always the chance of a possibly fatal mistake.

overwhelming preference for males. The overall record shows 13.5 male victims for every female! If the data is limited to attacks along beaches the ratio drops to 9.1 males to 1 female. The old cliché about sharks being maneaters is almost literally true. But why is there this seeming preference for male victims? Casual head counts of beach populations show about equal numbers of males and females. Males do tend to be more active — swimming, diving and surfing — and they generally venture further from the shore. A possible related statistic is drowning deaths. It is the case, in the United States of America at least, that with drowning deaths in the range of ages that includes most shark attack victims (15-25 years), the ratio of males to females is 10.6 to 1. The similarities between these two ratios implies a relationship between the circumstances leading to drowning and shark attacks.

Is a person's race a factor in shark attack? All races have been attacked by sharks and there is no evidence that one race is more likely to be attacked than any other.

What about colour of clothing and bathing suits? Again there is no conclusive proof from the files that the colour of clothing was a factor in attacks. Laboratory tests with captive and wild sharks have shown that sharks are much more likely to approach bright, reflective objects for food than dark, dull ones.

Where do attacks occur? Shark attacks have occurred in every type of location where people go into the water and where sharks live. There are even reports of sharks stranding themselves in attempts to get at beach strollers. A large proportion of attacks occur at bathing beaches, in water less than two metres (79 in) deep, and within 65 m (213 ft) of the shore. This is because the largest number of bathers swim there.

Sharks generally prefer deeper water, only coming close to shore at night. About one-third of beach attacks occur at distances greater than 65 m (213 ft) from the shore, and the number of attacks is relatively independent of distance beyond that point. This can be explained if it is assumed that, while the number of people present decreases with distance, the number of sharks increases, thus the chance of an individual being attacked increases with distance from the beach. The chance of drowning also increases, since a swimmer is getting further from potential rescue. This is the probable reason for the high proportion of males being attacked. In observations made of swimmers at beaches, males are generally in deeper water than females, making their chance of encountering a shark that much greater. In fact the ratio of males to females attacked at

distances greater than 65 m (213 ft) from the shore is 31 to 1, much greater than the overall ratio.

Are divers with air tanks and skin divers safer from attack than surface swimmers? It has been widely thought that divers have the advantage of being able to see an attacking shark and take evasive action. But only half the divers attacked reported seeing the shark before they were hit. About 10 per cent of reported attacks are on divers. Since the number of divers in the water at any one time must be much smaller than 10 per cent of beach bathers, the odds of being attacked must be significantly greater for divers. No doubt diving activities such as spear fishing and collecting abalone in turbid waters, in places where sharks feed on seals, add greatly to the chance of attack. The chance of encountering a shark in the deeper water frequented by divers is also much greater. There is

WARNING FOR THE WARY

NORMAL BEHAVIOUR

Side

Front

THREAT BEHAVIOUR

Lifted snout

Hunched back

Side

Lowered pectoral fins

Front

Body language among sharks has been little studied yet, simply because there are not enough people working with these animals. The value of such research, particularly to divers, is illustrated by the discovery of a threat display exhibited by grey reef sharks *Carcharhinus amblyrhynchos* in the Pacific Ocean (although not in those found around Australia). A solitary shark will sometimes confront a diver with its back hunched, its nose turned up and its pectoral fins pointed down at an abnormal angle. If the diver continues to approach the shark, he or she risks a quick, slashing attack before the shark flees. During experiments conducted by United States' researchers these sharks were even provoked into attacking a small research submarine.

evidence that one species of grey reef shark *Carcharhinus amblyrhynchos*, found around many reefs in the Pacific Ocean, is territorial and becomes aggressive towards divers if it feels threatened by the invasion of its domain. These sharks go through a threatening display of back-hunching and erratic swimming before attacking, so alert divers are forewarned.

What is the relationship of water temperature to the chance of attack? More than 20 years ago two scientists, Dr V. M. Coppelson and Dr D. H. Davis, suggested that there was a strong relationship between the chance of shark attack and water temperature. The chance of shark attack below a water temperature of 20°C (68°F) was supposed to be low. Taken at face value the data from the file supported this view. Almost 80 per cent of recorded attacks occurred at temperatures above 20°C (68°F). However, David Baldridge put forward the very reasonable and logical argument that the relationship between temperature and attacks is due more to the comfort and physiology of people than to shark behaviour. Studies of the ability of people to maintain their body temperature in cold water show that in water below 20°C (68°F) the average person cannot maintain body temperature, and within a few hours will probably die of exposure. In census counts made at two popular beaches it was found that only an average of nine per cent of those present would swim in water colder than 20°C (68°F) at a given time. This is indeed fortunate, because at one beach, in Florida, the shark population is known to be much greater at temperatures below 20°C (68°F) than at the more comfortable (for humans) water temperatures above 24°C (70°F). The conclusion to the argument is that the probability of attack is greater in warm water,

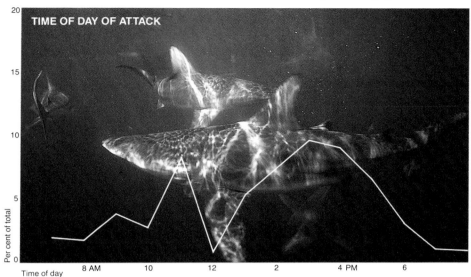

TIME OF DAY OF ATTACK

Per cent of total

Time of day — 8 AM — 10 — 12 — 2 — 4 PM — 6

DAY OF WEEK OF ATTACK

Per cent of total

Day of week — Sat. — Sun. — Mon. — Tue. — Wed. — Thu. — Fri.

COUNTRY OF ATTACK

Rest of the world 34 per cent of attacks

South Africa 8 per cent of attacks

Pacific Islands 12 per cent of attacks

United States of America 19 per cent of attacks

Australia 27 per cent of attacks

A whitetip reef shark among schools of fish on Australia's Great Barrier Reef. Sharks are obviously attracted to fish – particularly to large schools.

because the very large increase in bathers far exceeds the decrease in shark population. The chance of an individual being attacked in warm water is actually lower than in colder water. Wet suits, which allow people to go into colder water, and stay longer, are probably a factor in the relatively high incidence of attacks on divers, as well as the fact that the chance of encountering a shark is greater in deep water.

What days of the week and times of the day do attacks occur most often? By now it has been well established that there is a strong relationship between the number of bathers and chance of attack. It is therefore not at all surprising that more attacks take place on weekends than weekdays. In fact 65 per cent more attacks occur on weekends than weekdays. The time of day at which most attacks occur

is also strongly related to beach population. After dawn the number of attacks increases to a peak at about 11 am, drops to a minimum at noon, increases again to a broad peak in mid afternoon, between 2 pm and 4 pm, then decreases to another minimum after 6 pm. Baldridge found a similar pattern for beach population.

What circumstances may increase the chance of an attack? The presence of large numbers of fish, or fish behaving in an unusual manner, has been reported preceding many attacks. About 20 per cent of shark attack victims were associated with people pole fishing – they were either fishing themselves, netting fish, holding caught fish or simply standing or swimming nearby. In another 20 per cent of cases people were spear-fishing in the area of an attack. A

comparison of the number of people swimming to those fishing and spear fishing seems to show that those two pastimes have by far the highest risk of inducing an attack. Sharks are well known to frequent areas where refuse is dumped into the ocean. The presence of such material was strongly suspected to have been an important factor in about 25 per cent of the cases.

What are the characteristics of shark attacks? Individual sharks acting alone are believed to be responsible for 94 per cent of attacks. There was more than one shark involved in only 6 per cent of reported attacks. The shark was seen prior to the attack in only one-third of the cases. Practically all attacks were direct strikes on the victim. It was seldom that close passes were made before the attack, and in the majority of cases there

Humans are bitten by sharks mostly on the arms, hands, legs and feet. This is not surprising, because most attacks take place in waist-deep water, and the victim often tries to fend the shark off with the hands, if at all possible. The extent of wounds range from severe lacerations, which are the most frequently described injury, to the body being skeletonised in a small number of cases. Only on very few occasions has the victim been swallowed whole.

Head 2.4 per cent of attacks

Shoulders 3.8 per cent

Hands and fingers 20 per cent

Arms 23 per cent

Chest 5.5 per cent

Abdomen, stomach and waist 11.2 per cent

Back 3.7 per cent

Genitals 2.2 per cent

Buttocks 10 per cent

Thighs 33 per cent

Calves and knees 40 per cent

(Figures total more than 100 per cent because more than one body part was injured in some attacks.)

Feet and toes 23 per cent

THE DEADLY TRIUMVIRATE

Three species of sharks – the bull shark *Carcharhinus leucas* (top left), the tiger shark *Galeocerdo cuvier* (top right) and the great white shark or white pointer *Carcharodon carcharias* (bottom) – have been blamed for most attacks on humans. These three species were identified as being responsible for 80 attacks between them in the Shark Attack File, with the great white (32 attacks) as the clear leader. Other species classed as dangerous are the blacktip, blacktip reef, blue, Caribbean reef, copper, dusky, Galapagos, grey reef, lemon, nurse, oceanic whitetip, sand tiger, shortfin mako, sharptooth lemon, spinner, spotted wobbegong, tasseled wobbegong and tawny nurse sharks. All of these have actually attacked people or boats.

was only one strike. Few attacks involved numerous bites. This indicates that in many cases the attacking shark mistook the victim for a more usual kind of food, and did not attack any further when the error was discovered. It is fortunate that sharks, in most cases, do not consider humans to be suitable food. This information also refutes the long-standing notion that fresh human blood is a powerful attractant that excites sharks into a feeding frenzy. If this were so, the presence of blood would certainly have induced the attacking shark to strike the victim repeatedly. Most wounds occur on the appendages – the hands, arms, legs and feet. Lacerations of varying severity are the most common types of injury. About 25 per cent of attacks kill the victim. The most usual cause of death is shock, combined with a severe loss of blood.

What species of shark are most likely to be involved in an attack? In about one-quarter of the cases in the Shark Attack File the attacker was identified with varying degrees of certainty. Not too surprisingly the great white shark was blamed most often, with 32 attacks. The tiger shark *Galeocerdo cuvier* was

second with 27 attacks and the bull shark *Carcharhinus leucas* was third with 21 attacks. The rest are divided among about 30 other species known to attack people. The size of the attackers varied from under 500 mm (20 in) to over seven metres (23 ft) in length, with the average length being about two metres (6.6 ft). This is also about the average length of sharks caught.

What is the actual risk of being attacked by a shark? Other than saying that the risks are extremely low, it is very difficult to make quantitative estimates of the risk of attack. The true statement is often made that more people are killed each year by lightning than by shark attack. In the United States of America about 500 people are struck by lightning in a year. But the United States has a much greater incidence of lightning strikes than any other large country, and not everyone who is a potential shark attack victim is exposed to lightning, and vice versa. The only meaningful comparison that can be made is that between drowning deaths and shark fatalities along bathing beaches. There are at least some common factors among the victims in both

cases. The male to female ratio among victims is almost the same, being about nine or ten to one. In both cases the victims' ages have been in roughly the same range, between 15 and 25 years. If only drowning deaths along beaches where shark attacks have occurred are considered, together with the number of shark attacks in one year, it is possible to estimate the relative chance of drowning compared to that of being killed by a shark. For the United States including Hawaii, the chance of drowning is more than 1000 times greater than that of dying from a shark attack. Assuming all other factors to be the same, and taking into account the different population sizes and number of shark attacks, for Australia the chances are 50 to 1, and in South Africa 600 to 1. While these numbers indicate that the chances of being killed by sharks in Australia are considerably greater than in the other two countries, the chances are still really very small.

REDUCING THE RISK

Swimmers and divers can reduce the chance of being attacked by following a few simple rules: Never swim in areas where sharks are known to be common. If the area is unfamiliar you can find this out by asking local residents. If possible swim at beaches where there are life guards present. Never enter the water where people are fishing, either from the beach or from inshore boats.

If there are a number of people in the water do not separate yourself from them. There is safety in numbers. Avoid swimming near deep channels, or where shallow water suddenly becomes deeper. Do not swim alone, or at dusk or after dark, when sharks are feeding actively and likely to be closer to the shore. Do not enter the water, or if in the water leave immediately, if large numbers of fish are seen, or if fish seem to be acting strangely. Be alert for unusual movements in the water. Do not wear a watch or other jewellery that shines and reflects light. Do not enter the water with an open wound, and women should not swim during their menstrual periods.

IF THE WORST HAPPENS

Ms Beulah Davis, Director of the Natal Anti-Shark Measures Board in South Africa, in collaboration with two physicians, has published detailed instructions for the treatment of shark attack victims. Their advice is summarised below.

Shark attack victims usually die from a combination of shock and blood loss. Therefore, the following things should be done as soon as possible:
● Remove the victim from the water as quickly as possible and place him or her head-downward on the beach slope to combat shock by increasing blood flow to the head.
● Control bleeding by pressing on pressure points, or by applying tourniquets. Efforts to stop bleeding should start while the victim is still in the water.
● Notify a doctor, paramedic or hospital. Take the victim's blood pressure and pulse

rate if possible for future reference by a doctor or hospital.
● Do not give the victim warm drinks or alcohol, only sips of fresh water. Protect the victim from cold by wrapping him or her in a blanket to minimise heat loss.
● Bring aid to the victim rather than take the victim to the aid. This is because movement can increase shock. The victim should not be moved unless he or she has recovered from shock and a doctor is present. Victims are better off left alone than moved unwisely or unnecessarily.
● Untrained people should not try to help the victim in any way, other than by carrying out the steps outlined above – more harm than good can result from well-meant but incorrect attempts to render aid. Experts think that this is one of the most important factors in determining whether a victim survives or not.

TESTING A NEW THEORY

In the early 1970s an American scientist suggested that sharks were frightened of divers wearing black and white striped wet suits because they resembled giant sea snakes. This exciting possibility was tested some years later near Osprey Reef in the Coral Sea by diver Valerie Taylor.

A second, narrow-banded, suit (below) was made and tried out on reef sharks after the broad-banded suit (bottom right) had failed. This second suit seemed to have no deterrent effect on sharks either.

'I was sceptical from the first about claims that sharks are frightened of banded sea snakes. The initial experiments had apparently been carried out around Lord Howe Island where sea snakes are rarely seen. On several occasions I had seen both sharks and banded sea snakes attracted to baits, but had never noticed any obvious animosity between them.

'I had dived with the three species of sharks that we encountered that day on Osprey Reef on many occasions. All could be tempted to approach a diver with an offer of food, but they were always cautious, and would never come close enough to touch. The banded suit, however, seemed to change all that.

'Three of us dived on the reef and positioned ourselves on a ledge about 25 m [80 ft] beneath the surface. Only I was wearing a banded suit. I opened the bait bag, sending blood billowing into the water. Immediately a whitetip reef shark [*Triaenodon obesus*] appeared from behind and pushed into me. It was followed by two more sharks

that pushed into my legs so hard that I could not open the bag to extract more bait. I held one steady by putting a flipper on its head, and wrestled more tuna from the bag.

'This violent tug-of-war provoked immediate action from the larger sharks further out. Eight grey reef sharks [*Carcharhinus amblyrhynchos*] torpedoed in. It was as though I did not exist. One moray eel, two grouper and dozens of other fish added to an incredible muddle. I snatched back the bait bag twice, tearing it from the mouths of sharks as they tried to drag it away. A bump in the side had me spinning away from a grey reef shark, which had charged in at my feet.

'More sharks joined the action. A whitetip forced its head between my legs from behind. I jammed them shut, ramming my (blunt) knife onto its head without effect. The small shark struggled forward, grabbing at the bag. As I wrestled with this shark a second hard, grey head pushed under my arm. I could feel the suit catching and stretching on their skins as the sharks jostled about me. No baits were visible, but

every shark knew they were there, and was trying to reach them. Then, as suddenly as it had started, the action stopped. Only an eel and a few stray whitetips remained.

'Further out the cat-eyed greys patrolled back and forth, sleek shapes in the deep blue ocean. A fat 2.5-m (8.2-ft) silvertip shark [*Carcharhinus albimarginatus*] moved in, but for all its size and speed it lacked the courage of its brothers and sisters, and did not come closer. I sensed that the other divers were the deterrent, not me.

'Six months later at Great Detached Reef, further north, I tried again. I had been told that the stripes on my first suit may have been too wide, so I had had a second suit made with narrower bands. The results, however, were the same.

'After fairly extensive tests I feel sure that black and white striped suits do not deter sharks at all. In over 30 years of diving I have never had to push grey reef sharks away. It was only when I was wearing the striped suit that they treated me without respect, almost as if they could not see me.'

Some banded sea snakes are highly venomous, and can be dangerous to humans. This observation led to the development of the theory which claims that sharks are frightened of banded sea snakes, and that divers wearing black and white striped wet suits should be safe from attack. Sceptics questioned the reasoning, pointing out that not all black and white sea snakes are venomous, and that some sharks have been known to eat sea snakes as part of their normal diet.

Wary, and clutching a knife for protection, Valerie Taylor watches as a grey reef shark and a whitetip close-in on her. Rather than deterring the sharks, as it was supposed to do, the striped suit actually seemed to make the sharks less cautious in her presence.

A diver despatches a mortally wounded grey nurse shark. Despite the lurid stories, sharks have far more reason to fear humans that humans have to fear sharks.

TALES OF TERROR AND BRAVERY

There is little to say about the moment of crisis in an attack – the actual bite. Sharks are efficient predators with sharp teeth and powerful jaws – the outcome of an attack is a foregone conclusion. The real interest revolves around the amazing feats performed by humans in extreme circumstances, as they cling desperately to life, and the bravery of those who go to their aid.

A near-straight five-kilometre (three-mile) stretch of sand from Aldinga Beach to Cactus Canyon is a highway on the South Australian coast for cars towing boat trailers; a boat-ramp sign puts the 'beach speed limit' at 25 km/h (15 mph). Off Snapper Point, a wide sea-level reef pushes out into an aquatic reserve: fishing is prohibited.

Fishing was not prohibited on Sunday 12 March 1961 when Brian Rodger took to the water. Rodger, 21, his body trimmed by a summer of exercise and diet, was competing in the annual spearfishing competition conducted by the Cuda Spearfishing Club and the Underwater and Photographic Association, of which he was president.

After four hours in the water, Rodger had an impressive take: he was missing only a herring kale to complete a catch of all the common species. He headed for a deep ledge one kilometre (1100 yds) off the sand highway, speared two kale and a morwong and, feeling well pleased with himself, headed back to shore.

As two huge kingfish passed beneath him, he thought idly that the sight was unusual, perhaps he might even sight a big shark. He had never seen a really big shark, and marked it as a gap in his diving experience. (Everything is relative: the week before, Rodger had seen a 2.7-m (8.9-ft) whaler shark, and a companion had lost 23 kg (51 lb)

Diver Valerie Taylor rests on the deck of a boat while she waits to be taken to hospital for treatment after a shark bite. In over 30 years of diving – much of the time with large sharks – she has only ever received minor injuries.

Philip Horley, aged 17, points to damage caused to his surfboard by a great white shark. The incident occurred at Cactus Beach, Western Australia, in August 1977 when Horley was surfing with friends about 270 m (300 yds) offshore. He was flung from his board by the shark and was lucky to escape with severe gashes to his thigh and knee. It is surprising, considering the popularity of surfboards and the fact that surfers take pleasure in ignoring the danger, that more riders are not taken by sharks.

A Javanese native being harrassed by a strange furry shark. The artist seems to be in two minds as to whether he is illustrating a shark or a seal. Many species of sharks, including all the dangerous ones, are found throughout the tropical regions of the earth. The native inhabitants came to terms with their presence long ago. However, despite the worship of shark gods, and elaborate rituals designed to ward off attack, fishermen were regularly attacked, and sometimes killed.

of fish and a pair of plastic shoes from his float. Rodger was mildly irritated: they were his shoes.)

He breathed deeply through his snorkel, preparing to dive after the kingfish. Suddenly something sharp and ragged seized his leg and hip and shook him; he twisted and saw a 3.6-m (11.8-ft) great white shark. As his body contracted in a spasm of pain and fear, his left arm jabbed for the shark's black eye. He missed. The teeth on the shark's upper jaw slashed his left arm to the bone. Unaccountably, the shark released him.

Rodger, superbly fit, at home in the water, and psychologically comfortable with the risks of shark attack – as all experienced divers are – did not panic. He was a competitive killer of fish, and as the shark turned around him in a tight, fast circle, the thought struck him

that he would have to be good to get out of this. For the first time he saw the shark's full size and sensed its enormous power.

He speared the shark on the top of its head, about 70 mm (2.8 in) behind the eye. It stopped its charge – and shook out the 1.5-m (4.9-ft) stainless steel spear. Rodger, badly hurt and bleeding profusely, felt elated. 'It was quite irrational, but for a moment [the spear hit] was all that mattered.'

Then the spade tail flicked away and Rodger, suddenly alone, realised how desperate things were. He was more than 700 m (765 yds) from shore. His leg and arm were cut to the bone, pumping inky clouds of blood into the water.

Tentatively, he tested his lacerated leg and found to his surprise that he could still move it. His vital organs were untouched: if

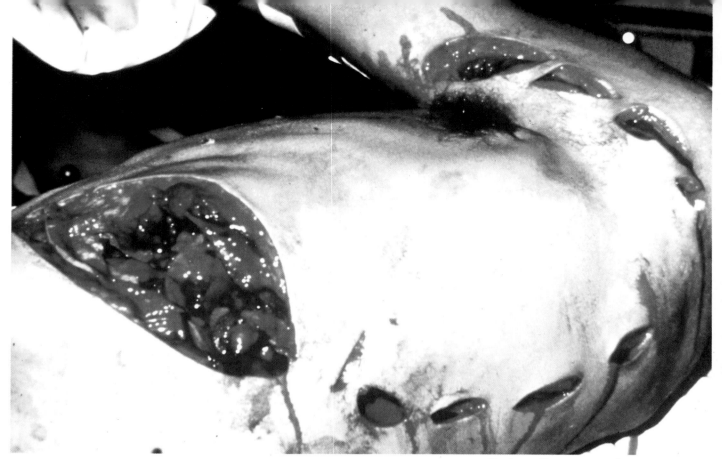

Rodney Fox, an award-winning Australian diver, must be one of very few people to have experienced and survived the bite of a large shark. The attack exposed his stomach, lungs and rib cage; the flesh on his arm was stripped to the bone and his ribs were crushed – one of them puncturing his lung. Fox spent four hours on the operating table in Adelaide (above) and he was lucky to be attended by a surgeon who had just returned from a course in England on chest operations. The wounds needed 462 stitches. Not long after the attack, Fox was back diving again, and a year later he was part of the team which won the Australian national championship. He will carry the scars (left) of his awful encounter for the rest of his life.

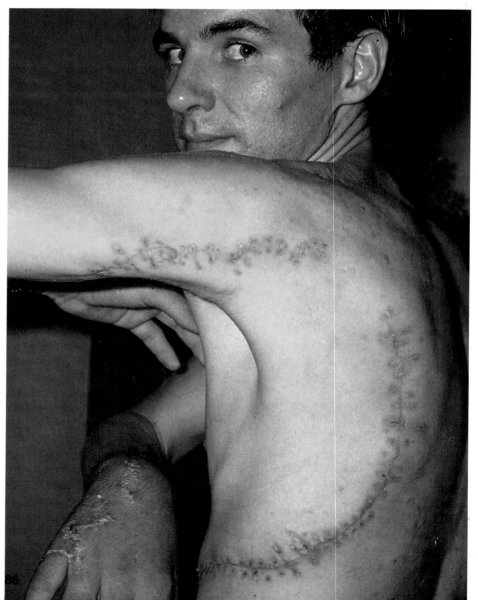

he could reach the shore he could survive. He began the agonising swim, worrying about his loss of blood, then realising he could make a tourniquet from his spear-gun rubber. He twisted the rubber tight around his upper thigh, using his knife as a windlass, and secured the knife handle under the bottom edge of his wet suit jacket.

The tourniquet reduced the blood flow, but Rodger felt his strength draining dangerously. He released his speargun, lead belt and the float with all his fish, bitterly regretting it. He thought later how absurd it was, longing to keep a few dollars worth of equipment and a

HEADLONG INTO DANGER

Two pearl divers have had miraculous escapes in strangely similar attacks. In both cases the shark grabbed the diver by his head, and was forced to let go when the victim struggled and gouged at the shark's eyes. The first incident took place in about 1913 in Torres Strait, between New Guinea and Australia, with the second 24 years later in the same area.

The first victim was a young Thursday Islander named Treacle. He dived straight down from a pearling boat into the open jaws of a large tiger shark, which almost succeeded in tearing his head from his body. Deep gashes in Treacle's neck exposed his jugular vein and his head and shoulders were also badly cut.

The 38-year-old victim of the 1937 attack, Iona Asai, recounted his experience: 'First time I went down I found one pearl shell and put it on the deck. Then I went down again

the second time and found another pearl shell. The third time I dive and walked on the bottom. I was behind a little high place, the shark was on the other side. I never saw him and he never saw me. I saw a stone like a pearl shell on the north side, and when I turned I saw the shark six feet [1.8 m] away from me. He opened his mouth, already I have no chance of escape from him. Then he came and bite me on the head. He felt it was too strong to swallow and put his teeth around my neck. Then he bite me and I felt his teeth go into my flesh. I put my hands around his head and squeeze his eyes until he let go me, and I make for the boat. The captain pulled me into the boat and I faint. They get some medicines from a school teacher.'

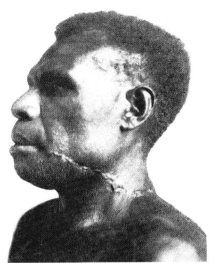

***Iona Asai survived** an attempt made by a shark to bite his head off.*

few fish for points in a competition when he might die, but he had to force himself to let them go.

The swim seemed to take forever. He found breathing through the snorkel difficult and turned on his back. He raised his arm, waved and shouted 'shark' but the families on the beach did not react. He subsided, cursing the waste of energy, when a small rowing boat manned by two furiously paddling spearfishermen approached. It was obvious the two-metre (6.6-ft) boat would not hold the three of them; one jumped into the bloodied water, helped to heave Rodger into the boat, then swam behind to push it.

A dozen divers ran across the reef, lifted the boat from the water and carried it to shore, where a St John's Ambulance man tended Rodger while the divers prepared an old door as a stretcher and carried him up the steep rise backing the beach to the ambulance. Police escorted the ambulance in the 55-km (34-mile) dash to the Royal Adelaide Hospital.

Doctors inserted 200 stitches in Rodger's wounds in a three-hour operation. He had lost four litres (7 pints) of blood. He resumed skindiving less than three months

later, and before the end of the year set a new Australian record by reaching a depth of 45.4 m (148.9 ft) in a lake without air tanks.

While Brian Rodger fought off the great white, Rodney Fox, another spearfishing competitor, cruised the bay nearby. Fox knew nothing of the drama until he came ashore – but he too had a brush with a great white shark, a huge brute which circled him ominously and occasionally came so close he could have reached it with his speargun. Fox repeatedly dived to the bottom, moving cautiously shorewards and after 10 minutes the shark lost interest and moved off. Later, when he talked about the attack with Rodger, they agreed that all you could do with a shark was go for the eyes, that was the only vulnerable spot.

Two years and eight months later, on 8 December 1963, Fox was competing in the South Australian Spearfishing Championships, having won the title the previous year. The scene was again Aldinga Bay – although this time the divers were not using fish floats: competition boats picked up the catches.

Fox was in superb form, drifting, gliding, spearing his quick

elusive targets with the practised ease of a born competitor. With an hour left, he looked likely to win the title again. He was one kilometre (1100 yds) offshore, drifting in for a shot at a dusky morwong, sure of the kill, his finger tensing on the trigger, when something huge hit his left side – 'it was like being hit by a train' – knocking the gun from his hand and tearing the mask from his face. His next impression was of speed, surging through the water faster than he had ever done, a gurgling roar in his ears, and of the easy, rhythmical power of the shark, holding him as a dog does a bone.

With his right arm he clawed for the shark's eyes; it released its grip and Fox instinctively thrust out his right arm to ward it off. The arm disappeared into the shark's mouth, lacerating the underside on the bottom row of teeth. As the horrified Fox jerked it out, the arm caught the upper jaw. In extremity men do amazing things: Fox, terrified of the open maw, tried to bear-hug the shark, to wrap his arms and legs around the abrasive skin, to get a purchase away from the teeth. It did not work – the

Continued on page 90

THE PERSISTENT SHARK

Few of the victims of shark attacks get more than a fleeting glimpse of their assailant. However, in February 1966 – in a unique incident – a young Sydney boy was badly mauled by a shark that refused to release him, even when it was carried ashore.

It was a warm Sunday afternoon, typical of late summer. At the small seaside town of Coledale, 64 km (40 miles) south of Sydney, a crowd of holidaymakers were enjoying the last of a fine weekend. From the beach, lifesavers watched as about 60 people swam, paddled and played in the stretch of warm, shallow water between the patrol flags. Just outside the flags a few surfboard riders made the best of the day's small waves.

About 27 m (30 yds) offshore, still only in chest-deep water and well within the patrolled area, 13-year-old Raymond Short was swimming just inside a sandbar which ran parallel with the beach. He was down from Sydney with his parents for the weekend, staying in a caravan park near the waterfront.

None of the swimmers in the water, Raymond included, took much notice of the patches of weed drifting in with the tide. There had been weed in the water – which was slightly cloudy – for the past couple of days. One of the lifesavers on duty that day, Eddie Patmore, remembers glancing idly at what looked like a particularly large clump drifting among swimmers and heading towards Short, but he, too, ignored it.

Suddenly the afternoon peace was shattered as Short screamed out in panic, shouting for help at the top of his voice, as the water around him turned pink with blood.

At first lifesavers thought that Short had become entangled in seaweed, but as the swell rose they could see a shark wallowing in the water beside him. Immediately the shark alarm was sounded and six lifesavers dashed into the sea, struggling to run through the water as they converged on the boy.

When they reached Short he was in chest deep water beside the sandbar. The shark was nowhere to be seen, so they started pulling him towards the beach, hampered by the small waves that were breaking over them. They were quickly joined by a nearby surfer who offered his board as a stretcher.

As the lifesavers half-lifted, half-dragged Short through the water he kept screaming out 'It's still got me. Get it off. Get it off.'

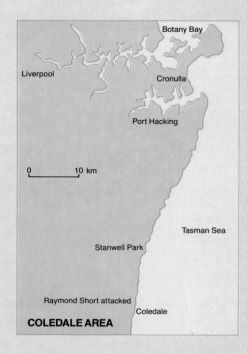

One of the Coledale lifesavers who helped in Raymond Short's rescue holds open the mouth of the attacking shark. A healthy great white of this size could easily kill an adult human.

COLEDALE AREA

Incredulous, one of the lifesavers, Raymond Joyce, thrust his hand into the murky water. To his horror, his fingers touched the snout of a shark, its jaws still firmly clamped onto the boy's leg.

Four more lifesavers now joined the struggling group, and as two continued to help Short towards the beach, the rest lifted the 2.4-m (8-ft) shark bodily from the water and carried it ashore.

Near the beach the shark still stubbornly refused to open its jaws. In desperation one of the lifesavers grabbed a nearby surfboard and smashed it down repeatedly on the shark's head, but without any noticeable effect. Another lifesaver, who lived nearby, rushed home to fetch a rifle, and while he was gone the rescuers renewed their

Raymond Short (fourth from left), and the seven Coledale lifesavers involved in his rescue, at Government House in Sydney on 18 November 1966. The men, who all received the Queen's Commendation for brave conduct, are (l to r) Raymond Robertson, Clarence Taylor, Brian Joyce, (R. Short), Warren Haberley, Dallas Haberley, Lessley Kennedy and Raymond Joyce.

desperate assault on the shark, which finally released its grip.

Immediately towels were packed around Short's lacerated legs, and he was rushed from the beach to a nearby hospital.

The boy was in the operating theatre for two hours while doctors repaired the damage to his badly mauled legs and hands. He received a massive transfusion of blood, and was in a critical condition when he left the operating theatre.

The shark was identified as a female great white, which weighed about 136 kg (300 lb). A close inspection showed that it had long, deep gashes down its sides near its tail, as well as numerous teeth marks elsewhere on its body. The wounds were only partially healed, and experts later suggested that the shark may have been weak and unable to catch its usual food. In desperation it had attacked a swimmer instead.

Doctors in the hospital watched anxiously as Short's condition gradually improved over the next couple of days. At one stage it had seemed possible that his right leg might have to be amputated, but this was not necessary. A few days later he returned home, and eventually recovered completely.

The jaws were removed from the shark, cleaned and offered to Raymond as a memento of his brush with death. Instead he insisted that they be presented to Coledale Surf Life Saving Club, as a tribute to the bravery of its members.

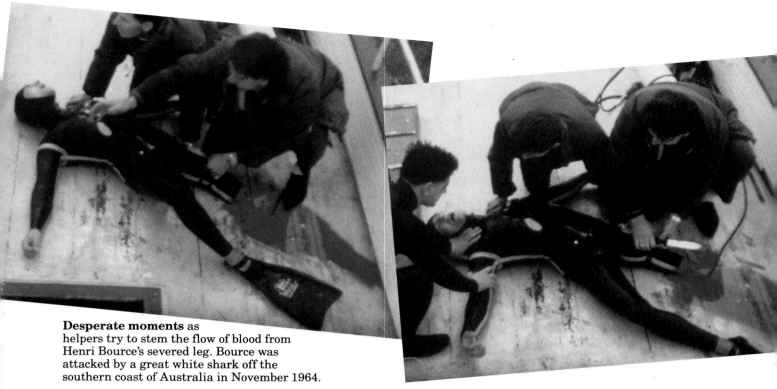

Desperate moments as helpers try to stem the flow of blood from Henri Bource's severed leg. Bource was attacked by a great white shark off the southern coast of Australia in November 1964.

shark was too big for him to hug.

He suddenly realised another need even more urgent than fending off the shark – air. He pushed away, kicked for the surface, gulped one breath and looked down on a scene that burnt itself into his memory. His mask gone, his vision blurred, he floated in a pink sea, and a few metres away was a pointed nose, and a mouth lined with razor sharp teeth, coming at him.

In desperation, Fox kicked with all his force at the shark. It was a terminal gesture, pointless, useless – but it worked: the shark turned from Fox, lunged for the buoy tied to his belt, swallowed it whole, then plunged for the deep. Fox, his ears roaring, reached for the quick-release clip on his belt. He could not find it. He realised the shark must have wrenched the belt around his body: the clip must be at his back. His lungs drained of air, his mind becoming fuzzy, he thought: that's it.

Then the impossible happened: the buoy rope snapped. Fox realised later that the shark must have bitten the rope when it attacked him. He floated to the surface, where his friend Bruce Farley and another man who had seen blood in the water pulled alongside in a boat. Fox's arms were so lacerated he could not raise them, so his friends gripped his wet suit and rolled him into the boat. Blood was pouring from his wet suit. Farley ran the

boat onto the horseshoe reef. As they lifted Fox from the boat, loops of his intestines emerged from the hole in his belly; a bystander who had studied first aid for the police examinations pushed them back with his fingers. Fox was bundled into a car, straining to breathe – his left lung had collapsed. As the car sped for Adelaide, a friend, sick with anxiety, talked him on: 'You've got to keep breathing. Come on, keep trying, Rodney. Think of Kay and the baby. Keep going.' While his collapsed lung gurgled and he heaved the air into his chest, his most vivid sensation was of swaying in the back of the car as it reached 150 km/h (95 mph). An ambulance dashed to meet them, and Fox was in hospital in Adelaide within an hour of leaving the water.

Later, his wet suit removed, the extent of his wounds apparent, he heard someone suggesting that this man was near death, they should call a priest. He still had the energy, blood gurgling in his throat, to say: 'But I'm a Protestant!'

He recovered completely, and in 1964 Fox, Rodger and Farley won the Australian Spearfishing Championship team's event.

Attacks on skindivers generally follow a pattern: the strike is quick, the shark unseen, the diver's first impression being of the shark's weight and power. Then the instinctive reaction – gouging the shark's

eyes, kicking, anything to stay away from the teeth. In these first frantic seconds, divers commonly experience a curious detachment, fighting for their lives at one level, and with another part of themselves seeing the deadly encounter calmly and analytically. Because divers are physically at home in the sea and familiar with the presence of sharks, their reactions maximise their chances of survival. After the shark breaks off the attack, their ease of movement, ability to think rationally and above all their level of physical fitness are the factors that decide whether they reach shore alive or not.

With swimmers, bodysurfers and board riders the pattern is different. Most attacks take place in waist-deep water, and the first bite is usually to the most accessible parts of the body – the legs or feet. Many victims experience a bump or brush, then the brutal realisation that it is a shark, followed by a scream of fright, a frenzied struggle to escape and often, at the moment of safety, the first experience of pain and the discovery that a leg, foot or hand is missing.

Although most near-shore attacks take place in water only one to two metres (39 to 78 in) deep, some occur in water so shallow that the shark is in danger of being beached. In 1933 a man sitting in less than one metre (39 in) of water

Henri Bource, diving again after the attack that cost him his leg in 1964, holds open the mouth of a great white shark – the species that almost ended his life. He claims that he is not bitter about the incident, saying simply that: 'They [sharks] do what nature intended them to do …'

A native boy disappears in a flurry of water as he is seized by a shark. It is impossible to verify the authenticity of this photograph, taken in 1901, but the attitudes of the other people seem to suggest that it may be genuine.

at the mouth of Charleston Harbour in the United States of America, had his right knee seized by a 2.4-m (8-ft) lemon shark. When he lashed out with his left leg, that too was bitten. In 1961 a 13-year-old boy was standing on a submerged rock, also in water less than one metre (39 in) deep at Winklespruit beach near Durban, South Africa, when he felt something touch his right foot. He reached down with his right hand which was promptly bitten. Lifesavers carried him ashore, followed by a two-metre (6.6-ft) shark. His right foot was missing and his hand deeply bitten. An even more remarkable case occurred in 1966 when an 8-year-old boy and his mother were walking along a beach

in Florida. The boy was splashing around in about 30 cm (12 in) of water near the shore when his mother spotted a 'grey form' out of the corner of her eye. As she grabbed the boy, lifting him clear of the water, a shark rushed past so quickly that it was carried up on to the sand. It lay exposed for a few minutes before a wave carried it struggling back into the ocean. The pair continued their walk, now well back from the water's edge, and were followed by three or four sharks a short distance offshore.

Less than one kilometre (1100 yds) from the attack on the 13-year-old boy near Durban, Damon Kendrick, aged 15, reported a strange phenomenon when he was

attacked while bodysurfing after a lifesaving championship on 13 February 1974. A friend with him shouted a warning. As the shark bit into Kendrick's right leg, he heard 'a growl, best described as [the sound made by] a father who wants to amuse his child in a swimming bath, growling and blowing into the water at the same time'. Experts agree that sharks do not growl: perhaps what Kendrick heard was the commotion of the water and the shark's teeth grinding through his shinbone. The bite lasted two seconds and then a wave washed Kendrick ashore. Holding his injured leg up in front, he pushed himself backwards up the sloping beach, watching a 'river of blood'

The awesome size and power of the jaws of a great white shark are clearly conveyed in this view of an approaching monster. Even the bite of a two-metre (6.6-ft) shark has been measured at 3 tonnes per sq. cm (21 tons per sq. in).

companion towards three yachts and a powerboat moored 250 m (273 yds) off Fourth Beach, Clifton, South Africa. Halfway there, Spence stopped and, joking to his companion, re-enacted the struggles of the girl shark victim in the film *Jaws*. They then swam seawards for three minutes and trod water near the yachts. Spence later reported 'a hard thump on my side and I felt a vice clamp on my chest'. The shark quickly released its grip, then swam in front of him. Spence saw the open mouth as the shark appeared to be looking at him. Then its back and dorsal fin broke the surface and it disappeared. A dinghy quickly picked Spence up and took him ashore where doctors attended him. His wounds were superficial and he recovered completely. It seemed the shark, perhaps attracted by the vibrations caused by Spence's *Jaws* imitation, may have mistaken him for a fish in trouble.

Not all shark attack victims are as lucky as Geoffrey Spence. In about 35 per cent of attacks the victim dies – either as a direct result of the bite, later through blood loss, or from shock or drowning.

The most frequently reported serious injuries are the loss of hands, feet, arms or legs – in 26 per cent of attacks these were bitten off directly, or were so badly mauled that they had to be surgically removed later. In some seven per cent of reported attacks the body cavity was opened by a bite, and in just over two per cent the trunk was severed, or the victim was swallowed whole.

One rare eyewitness account of a man being almost eaten whole by a shark is that reported by Gerald Lehrer. In midsummer, 14 June 1959, Lehrer and his friend Robert L. Pamperin, were diving for abalone in La Jolla Cove, near San Diego in the United States of America. They were near underwater rocks, less than 100 m (110

run from his leg. 'I have never seen so much blood before, and what scared me most was that it all came from me,' he said later. He survived, although his leg had to be amputated below the knee.

Five weeks later James Gurr was riding a surfboard just south of Inyoni Rocks, also on the Natal coast of South Africa, when he saw a shark's fin heading straight for him. He lifted his legs. The shark hit the surfboard, spilling him into the water. As he was remounting he felt a 'violent shove' from the shark under his arm. He panicked and paddled madly shorewards, feeling he had 'unbelievable strength'. Again he was bumped from his board by the shark,

but he remounted and paddled furiously as the shark zig-zagged in front of him. He caught a broken wave over the top of the shark, hit the beach, and sprinted away from the water. A friend later retrieved the board, which had teeth marks in the fibreglass.

The nature and outcome of many attacks seem to confirm the theory that at least some shark attacks are the result of mistaken identity. Once the shark realises that its human victim is not a seal or a large fish it loses interest and swims away. This appears to be what happened in the case of Geoffrey Spence.

On 27 November 1976 Spence, aged 19, was swimming with a

Many grisly remains have been removed from the stomachs of captured sharks. In 1886 this shark was caught in Sydney Harbour, and when cut open was found to contain several bones (including the lower portion of a human backbone), pieces of flesh, the leg of a pair of grey tweed trousers (the pocket of which still contained a penny) and the buckles belonging to a pair of braces. Simple detective work led to the trousers being identified. Their owner had not been seen since heavy seas had swamped his boat some days before the capture of the shark.

yds) from shore, and dozens of other swimmers were splashing in the water nearby.

Lehrer surfaced and looked towards Pamperin, a few metres away. Suddenly he saw a black dorsal fin slicing towards Pamperin. Lehrer screamed a warning and pointed to the fin. As he did so, Pamperin rose high in the water, as if he had stepped onto a submerged rock, then plunged back, threshing his arms. Lehrer, ducking his face beneath the surface, saw a shark about six metres (20 ft) long with Pamperin's upper body protruding from its jaws. Pamperin's mask was gone, his eyes were open and staring upwards. Lehrer could not tell if he was alive. He swam for shore and shouted a warning to the other bathers. Lifeguards in boats, a Coast Guard helicopter and divers

searched the area for hours, but no trace of Pamperin was ever found.

The attack on Pamperin gave him no chance at all of survival: a bite through the torso means either instant, or very quick death. Many survive bites to the arms or legs, although shock and blood loss are often fatal.

On 25 July 1936, 16-year-old Joseph Troy was swimming in Buzzards Bay, Massachusetts in the United States of America with Walter Stiles, an older man. They were about 130 m (143 yds) off-shore and about three metres (10 ft) apart. Suddenly both saw the familiar black fin alongside Troy. It disappeared, and then the shark struck the boy and dragged him under. Stiles dived and saw Troy, his left leg between the shark's teeth, kicking and struggling to break

ATTACKS THAT NEVER WERE

Such is the public fascination with sharks and shark attacks that any spectacular photographs of incidents involving people always have a ready market. It is inevitable, therefore, that enterprising divers are tempted to fake attacks.

In 1968 *Life* magazine ran a series of dramatic photographs in a story entitled 'Shark Kills a Diver'. In a sequence of five gory photographs a diver was shown being fatally bitten by an attacking great white. The story told how 32-year-old stuntman José Marco was acting out a scene involving a drugged bull shark when a great white broke through a net surrounding the underwater movie set and mauled him. Only later was it revealed that the photographs were fakes – somehow publicity stills had been circulated to various magazines as genuine shots.

In the 1960s a bizarre competition developed between two Australian underwater photographers as they attempted to outdo one another with dramatic photographs of underwater attacks. No sooner would one manage to have a (faked) photograph published of a diver fending off an attacking shark with a spear, than the other would immediately set about constructing an even more dramatic scene with a larger shark and a thinner spear. An imaginary diver, Ron Thomas, became the hero of these increasingly unlikely incidents, and for a time he was much sought after by television channels.

Intrepid diver Ron Thomas defends himself with a pitifully inadequate spear. It is fortunate that the shark in this faked photograph is dead.

Enterprising photographers were naturally willing to try and satisfy an insatiable demand for scenes of attacks – even if the sharks had to be dead before they would cooperate properly.

free. He did so and rose into Stiles' arms. Stiles, with Troy beneath one arm, sidestroked for shore. A man in a rowing boat who had seen the commotion and heard the screams picked them up – observing, as he did so, a large great white shark cruising about as if considering whether to make another attack.

Troy, parts of his left leg stripped to the bone, was rushed to a New Bedford hospital, but he died five hours later.

Another swimmer who, like Joseph Troy, had a heroic companion but was cruelly out of luck, was Albert Kogler.

On 7 May 1959, Kogler, aged 18, was treading water with a girl friend, Shirley O'Neill, 50 m (55 yds) off Bakers Beach outside San Francisco's Golden Gate. Suddenly he screamed and thrashed his arms,

and Shirley saw the upthrust of a monstrous tail. 'Go back! Go back!' Kogler screamed. Shirley hesitated. 'It's a shark – get out of here'. Shirley swam into the reddening water and reached for Kogler's arm. She found it barely hanging from his shoulder. Shirley put her arm around him as he floated on his back and both struggled shorewards, with Kogler kicking feebly. On the beach, his left arm almost detached, his neck and back deeply lacerated, Kogler lost consciousness. Two hours later he died in hospital.

Death came more rapidly to another youngster, 16-year-old Jeff Corner, while spearfishing with a friend, Allen Phillips, at Caracalinga Head on the South Australian coast on 10 December 1962. Corner was the state junior spearfishing champion and the pair

were about 180 m (197 yds) offshore while taking part in a competition. Phillips was pleased when, surfacing after a dive, he saw a commotion in the water around Corner, thinking 'Jeff's got a big one'.

Seconds later, he saw a large shark's tail break the water and thought 'it's probably pinching fish off the float'. He swam over to his friend and found himself in a cloud of blood. Sick with horror, he swam for their surf ski and paddled over. He tried to pull his friend onto the ski, but realised that the shark still had him in its grip. Corner disappeared beneath the ski and emerged on the other side. Phillips caught him by the shoulders and, still feeling the shark's grip, smashed at it with the ski paddles. Suddenly it let go. Phillips pulled

Continued on page 98

A Bolt
From The Blue

Most shark attacks are over before bystanders realise what has happened. Few people have the misfortune to see one at close quarters. Fate decreed that Lee Warner should watch, helpless, as a friend was savaged in the most terrible way.

Hundreds of kilometres of lonely coast stretch north from Perth, fringing one of the remotest and least populated parts of Australia. The few tiny settlements scattered along its length survive on fishing and a small seasonal influx of holiday visitors. Waves rolling on to its beaches from the Indian Ocean might have travelled unchecked across almost a third of the globe – the nearest land to the west is Africa, 8600 km (5300 miles) away.

In August 1967, on a bleak Saturday in winter, two divers arrived at Jurien Bay on this remote coast, 210 km (130 miles) north of Perth. They were Lee Warner, 26, a professional fisherman, and his 24-year-old companion Bob Bartle. Both were highly experienced divers and, as Western Australian State pairs spearfishing champions, they planned to enter the national championships to be held four months later at Busselton. The pair were at Jurien Bay that day for a minor competition that was to be held on the following Sunday. This

was to be the practice day, an attempt to find the best spots, and to assess local conditions.

From the top of North Head, parked among cars owned by other divers, they could look along the sand- and scrub-covered coast to the distant township of Jurien, 10 km (6 miles) away as the crow flies across the bay. Out to sea islands and reefs were picked out in white as winter waves broke around them.

The two men donned their wet suits and sorted out their equipment beside the car. Warner was wearing a full black wet suit, and Bartle a black suit with yellow seams and black flippers. Bartle's legs were bare, and both men wore hoods that covered their heads.

They entered the ocean below the headland and swam easily seaward into deepening water, heading for reefs about 1.6 km (one mile) offshore. The water was cold

Jurien Bay's brilliant, clear water looks benign on a calm morning. However, sharks – including great whites – frequent this area, and have been known to feed on seals nearby.

and slightly murky, although the weed-covered sea floor was visible about eight metres (26 ft) below.

After they had been swimming for about 20 minutes they reached a small dip in the sea bed about 700 m (760 yds) from the shore. The overhanging rocks fringing the depression looked like a promising spot for fish so Bartle dived down to check. There was nothing there, so Warner turned to continue seaward. Suddenly a huge, black shape shot

past him, just beneath his flippers. Warner later recounted the events of the next few moments:

'The shark came from the opposite way and went straight under me about 8 ft [2.4 m] down. It came out of the blue like a rocket and grabbed him [Bartle]. It moved so fast that by the time I looked back it had Bob in its mouth and was shaking him like a leaf. I rolled over immediately, dived and placed a spear in its head. It broke Bob in half and rose up at me with Bob's legs and flippers sticking out of its mouth. Bob's upper half floated to the surface. The shark began circling slowly. It made one pass at me, and I poked my spear gun in the direction of its eye. The gun struck behind its right eye, and a membrane appeared to cover its eye in a lateral plane. Realizing I was helpless, I retrieved Bob's gun which was floating near his body. As the shark passed by once more, I endeavoured to spear it in the eye. However, the spear passed over the shark. In his circling motion, he tangled this spear around Bob's float line and my spear line. I moved from the pool of blood and watched for some movement. The shark did not appear to be feeding. Bob's feet and flippers were still projecting from its mouth. The jaw must have been 2.5 ft [76 cm] wide. As there was nothing further that could be done, I swam towards shore...'

Warner now began a desperate swim back to the beach, 700 m (760 yds) away, glancing back constantly to check that the shark was not following him. As he struggled ashore he looked out to sea where the floats of other divers bobbed about on the surface, their owners unaware of the terrifying events that had just happened so close by.

Warner ran to the cars on the top of the headland and drove frantically down 8 km (5 miles) of bumpy track to Sandy Point, a tiny fishing settlement to the north. There he convinced Harry Holmes, a crayfisherman, to take him back to North Head in his boat.

Ninety minutes later, off the headland, they found the shark still tangled in the float line. An attempt was made to spear it again,

An aerial view of the Western Australian coast at Jurien Bay, north of Perth. Bob Bartle was taken 700 m (760 yds) off North Head, which juts into the ocean in the middle distance.

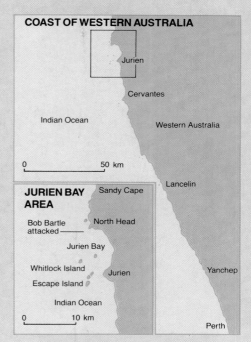

but it broke free and disappeared. They retrieved the upper half of Bob Bartle's body, which was floating nearby. It was unmarked, and an autopsy later revealed that there was still air in his lungs. Death must have been instantaneous.

Later, attempts were made to find out which species of shark had been responsible for the attack, but Lee Warner was unable to recall any details of its appearance. His entire attention had been riveted by the ghastly contents of its mouth. Experts thought that only a tiger or great white shark would have been large and powerful enough to have carried out such an attack. The fact that Warner saw a membrane cross the shark's eye seemed to rule out a great white – the most likely culprit – because they do not have a nictitating membrane. However, others have pointed out that the great white does roll its eye – displaying the white of the eyeball – just before it attacks. This is not, however, the question that is uppermost in Lee Warner's mind. To this day he still asks himself 'Why Bob?' Why did the shark pass him to attack Bartle?

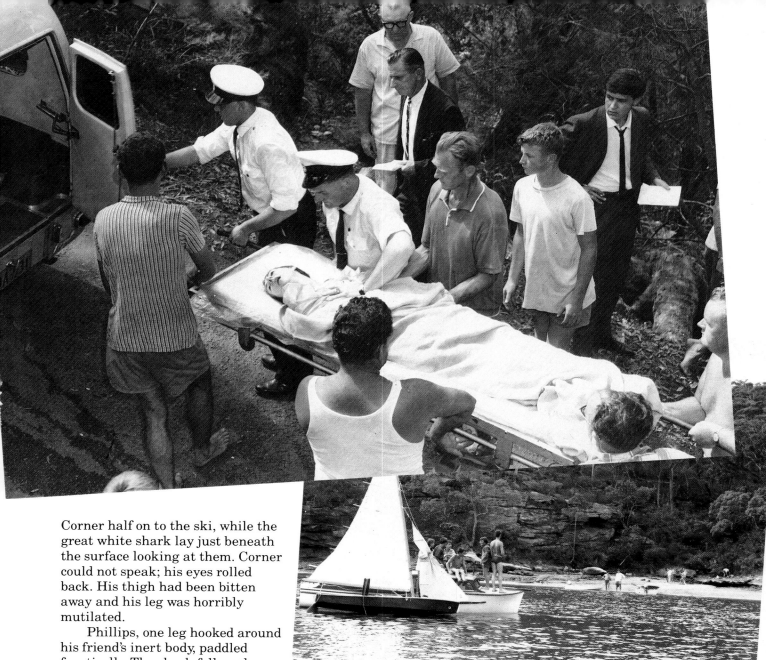

Corner half on to the ski, while the great white shark lay just beneath the surface looking at them. Corner could not speak; his eyes rolled back. His thigh had been bitten away and his leg was horribly mutilated.

Phillips, one leg hooked around his friend's inert body, paddled frantically. The shark followed. Another spearfisherman, Murray Brampton, paddled across and struck at the shark with his paddle.

Jeff Corner was dead when they reached the shore. His parents, and Phillips' wife and child, were among the hushed, horrified crowd.

Eyewitness accounts of shark attacks sometimes have the ring of authenticity – victims often report vivid details – but how much of the detail is true? Do eyewitnesses sometimes subconsciously persuade themselves that they have seen what they think they should have seen? Does terror play tricks on both perception and memory?

At Portsea, Victoria, on the Australian south coast, on 4 March 1956 six surf lifesavers went bodysurfing after a day's competition in surf rescue and swimming routines. All were young, fit and strong swimmers. Among them were John Wishart, 27, a plumber; Jack Hopper, 31, Captain of the Portsea Surf Life Saving Club; Gregory Warland, a sergeant at the Portsea Officers' Training School for the Australian Navy; and Richard Wright, 20. Two hundred metres (217 yds) off the beach, they trod water and waited for a wave.

This is Hopper's account of what happened next:

'Suddenly an enormous black shape coming from behind us darted between Wright and myself. He swirled in front of me close

Actress Marcia Hathaway (top) is carried to an ambulance shortly after being attacked by a shark on 28 January 1963. She died before reaching hospital from shock and loss of blood. The location of the attack, at Sugerloaf Bay (above) in Sydney Harbour, is notorious for sharks, and there have been five deaths near there since 1942. Miss Hathaway was with a number of other people wading in water only 76 cm (30 in) deep when a shark bit her below the calf. In a second lunge the shark almost tore her leg off as it sank its teeth into her upper thigh. Her horrified fiancé, who was standing beside her, fought the shark off and carried her to shore. The culprit was never found, despite an extensive hunt by fishermen.

The fate of this young girl, mauled during a boating trip from Trieste, present day Italy, in 1908, is not recorded. The Adriatic Sea holds the record for being the site of the world's most northerly recorded unprovoked attacks on humans.

enough to be touched and dived diagonally at John Wishart. Next instant there was a terrific crash as he took Wishart on the way down. The whole thing happened in a second. I imagined the brute brushed me with its tail, but it was the force of water as it swerved that hit me in the stomach.' He thought the shark was a whaler or tiger three to four metres (10-12 ft) long. Hopper swam madly for the shore.

This is Warland's account:

'Suddenly there was a splash and I could see the shark's jaws come out of the water. Then it seemed to splash down on top of Wishart. It was about four yards [3.6 m] away. There was swell in the water, then nothing. Hopper and myself first thought we could help, and tried to see Wishart. Then we

realised it was useless and swam madly for the shore.'

Here are two basically consistent accounts by eye-witnesses very close to the tragedy. Mrs G. Bell, however, was watching from a small bluff behind the beach where Wishart's mother and sister were sitting. Mrs Bell saw 'five of the men catch a big wave and ride it to the shore. But the poor sixth man missed it by about two feet [0.6 m]. He appeared to be waiting for the next breaker when he suddenly threw up his arms and disappeared. Then he began to thrash his arms in and out of the water. We then realised that he was being attacked by a shark. We saw a large dorsal fin break the water and a pool of blood spread over the surface ... The man in the water must have been

one of the bravest men that ever lived, the way he fought off the shark for four minutes.'

None of the accounts sound like verbatim tape transcripts, and it is possible some of the inconsistency arose from reporters condensing the eyewitnesses' interviews. Even so, how did Mrs Bell see a four-minute struggle by a lone surfer, when two surfers saw one man plucked from a line of surfers by a single strike?

Three days after the attack, a 227-kg (500-lb) shark swallowed a bullock's liver bait on a line from a floating oil drum. Fishermen in a 5.5-m (18-ft) skiff killed it with a twelve-bore shotgun and five bullets from a .22 rifle. Taken ashore at nearby Sorrento Beach, it was raised and disembowelled as a crowd of 1200 watched. The

stomach was empty. John Wishart's remains were never found.

With their wet suits for protection against grazing, facemasks that give clear underwater vision, flippers for speed, spearguns as weapons, and a psychological ease in the ocean, spearfishers have the best protection against sharks. Board and body surfers and beach swimmers are more vulnerable – with no weapons and no protective equipment – but at least they are usually close to the shore, to help in the water, and to medical attention on the beach.

Survivors of aircraft crashes or shipwrecks are by far the most vulnerable of all potential shark attack victims. They may enter the water shocked, concussed, with broken bones, and they may be bleeding. Land, and safety, may

be 1000 km away. They will have neither spearguns, nor masks, nor wet suits. If the sharks spare them, they may meet death by drowning, exposure, shock, starvation or a failure of the will to live. Common to all of them is the knowledge, rare in the most extreme human crises, that they are utterly helpless.

A United States' study of 2500 World War II air crash survivors' accounts revealed that there were only 38 shark sightings and 12 attacks, but the fear of shark attack was a common theme in every survivor's story.

Airmen surviving an ocean ditching were typically left with only inflatable lifejackets for support. They often discarded their heavy waterlogged boots so they could move easily, leaving their naked feet dangling like bait from beneath a lifejacket which prevented

them from diving or moving quickly if any sharks came.

United States Navy pilot Lieutenant A. G. Reading, with airman E. H. Almond, the radioman, ditched in the central Pacific Ocean in the mid-1940s, more than 100 km (62 miles) east of Wallis Island. Their reconnaissance plane had developed engine trouble. The impact of the crash knocked Reading unconscious. Almond pulled him from the sinking aircraft and inflated his lifejacket, ripping his own trousers off in the frantic struggle to get free of the cockpit. Reading recovered consciousness in the water.

Within 30 minutes they were being circled by sharks, although nothing happened for about an hour. Then they heard the sounds of aircraft. Reading reported later: 'I said to Almond, "Let's kick and

Gulf Stream, Winslow Homer's famous painting, raises many questions in viewer's minds. Homer is reported to have said: 'Tell them it is all right; the tornado does not hit him and he is picked up by the ship.'

This dramatic scene — which seems to owe something to Gericault's *Raft of the Medusa* — appeared on the cover of a Paris magazine in March 1906. The actual circumstances of the event it depicts are unknown, but presumably some of the sailors survived to tell the tale of their ordeal.

TERRIBLE DRAME EN MER. — NAUFRAGÉS ATTAQUÉS PAR DES REQUINS

splash around to see if we can't attract their [the aircraft's] attention." It failed, but suddenly Almond said he felt something strike his right foot and that it hurt. I told him to get on my back and keep his right foot out of the water, but before he could the sharks struck again and we were both jerked under the water for a second. I knew that we were in for it as there were more than five sharks and blood all around us. He showed me his legs, and not only did he have bites all over his right leg, but his left thigh was badly mauled. He

The predicament confronting these two unfortunate mariners will be familiar to a modern generation of comic readers. In 1860, when this engraving was published, artists often had to reconstruct scenes of events in farflung parts of the world from only the haziest of descriptions.

wasn't in any particular pain except every time they struck I knew it and felt the jerk. I finally grabbed my binoculars and started swinging at the passing sharks. It was a matter of seconds before they struck again. We both went under, and this time I found myself separated from Almond. I was also the recipient of a wallop across the cheekbone from one of the flaying tails of a shark.

Attacks on boats are common, and occasionally large pieces have been torn out of hulls and oars by sharks. The motives for such attacks are unclear, but it may be that sharks are attracted by flashing metal, or are simply curious and explore unfamiliar objects by biting them to see if they are edible. On one occasion the teeth of a giant shark were found embedded deep in the copper-clad hull of a large schooner when it was pulled from the water for routine maintenance. The strange looking shark attacking the rowing boat (above) dates from 1820, while the great white biting the stern of a power boat (left) was photographed in the early 1980s. The most terrifying incidents involve sharks that actually leap from the water into open boats. In December 1949 a 2.6-m (8.5-ft) grey nurse shark landed on top of a seasick fisherman who was resting in the bottom of a boat off Seaholme in southern Australia.

Almond's head was under water and his body jerked as the sharks struck it. As I drifted away ... sharks continually swam about and every now and then I could feel one with my foot.' At midnight Reading sighted a navy boat, called for help and was rescued unhurt. The fact that Almond lost his trousers as he pulled Reading from the cockpit may have made the fatal difference.

An Ecuadorian flight officer survived a similar tragedy in June 1941, after his plane ran out of fuel and he was also forced to ditch in the Pacific. The three survivors, supported by lifejackets, swam towards the nearest land through rough seas, but after five hours one man, weakened by the crash, the huge waves, and having swallowed a lot of salt water, died. The flight

officer, hoping to bring the body ashore, pushed the dead man ahead of him. A short time later something dragged the body from his grasp. After another five hours, the second survivor died in his lifejacket, and again the flight officer pushed the body ahead of him. Now, by the light of the moon, he saw dark shapes in the water and discovered that parts of the corpse's legs were missing. After two more strikes, the great black fish brushing against him as they bit into the corpse, he let it go and saw it quickly dragged away. At daylight he saw sharks following him closely. When he stopped to rest, his feet touched the backs of them swimming directly beneath him. But, although they followed him all that day, they did not attack

again. Thirty hours after the ditching he waded ashore. It was as if these sharks, on that occasion, would attack only the dead.

That was not always the pattern. At dawn off Guadalcanal, in the Solomon Islands, on a night in 1944, Lieutenant Commander Kabat, afloat in a lifejacket, felt a scratching sensation in his left foot and found it was gushing blood. A 'great fish' with a brown back was swimming away. It turned, breaking the surface in a steady line and came at the Lieutenant. He kicked and splashed in a frenzy and the great fish veered away. It came again. Kabat thrashed about and punched the shark, which again veered away, although he discovered it had taken a piece from his left hand. The shark attacked repeatedly at intervals of 10 or 15 minutes, biting Kabat's heel, elbow, hand and calf. But he found the salt water minimised the blood flow, and that he had no consciousness 'of great pain'. When he saw a passing ship and shouted for help, the shark bit his thigh to the bone. Men on the ship fired at the shark until Kabat screamed at them to stop, terrified that having survived repeated attacks he might be killed by a bullet within reach of safety. After months in hospital, he recovered.

Kabat's experience showed that even an utterly exposed and effectively helpless man could, against all odds, survive repeated attacks. Some survivors' experiences suggested that there was safety in sticking together. In September 1955 the four-man crew of a ditched DC-4 fought off sharks for nearly 44 hours between Wake and Johnston Islands in the Pacific by staying together in a square of marker dye and shark repellent. Although the repellent did not bother the sharks, the men stayed in a tight group slapping the water and yelling. Two of the four were not

wearing shoes and their feet were constantly bitten. Eventually one man, who lost a thumb to a shark, died of exhaustion and loss of blood. As a rescue ship approached, another lagged behind his two companions and was killed by repeated, aggressive attacks. It seemed that four men together cause enough noise and confusion to intimidate a shark pack which would kill an individual in minutes.

On another occasion, in July 1958, three men from the crew of a United States Air Force mail carrier survived for three days in shark-infested waters between Hawaii and Wake Island. They kept together beside a 'raft' of mail bags too small to support any one of them. When the sharks attacked, they all tried to climb onto the raft, kicking, splashing and screaming. Two other men who drifted away from the raft were killed.

Perhaps the worst loss of life to sharks in a single incident occurred when a German submarine torpedoed the British steamship *Nova Scotia* off the South African coast on 28 November 1942. The

Nova Scotia carried 900 men, 750 of them Italian prisoners of war. Giant sharks tore into the shrieking survivors in what one of them described as a 'feeding frenzy'. Clinging to rafts, oars and floating debris, the survivors remained for 67 hours in shark-infested waters. Eventually a Portuguese sloop picked up 192 men, the Portuguese sailors fending off sharks with boathooks during the rescue. There is no way of knowing how many of the 700 who died were shark victims, but the incident horrified the public in Britain.

In the United States of America, the Office of Scientific Research and Development, and later the United States Naval Research Laboratory, responded to this threat to morale in the allied navies and air forces, by conducting urgent research to develop a chemical shark repellent. Although a repellent was produced it was not effective and gave only peace of mind to those who used it. No effective repellent has ever been found, and work on the problem has virtually halted.

THE SAILOR'S REVENGE

From the earliest times sailors hated sharks and took pleasure in inflicting great suffering on any they caught.

In 1593 Sir Richard Hawkins — the first person to bring a shark back to England — made these observations during a voyage: 'Every day my company tooke more or lesse of them [sharks], not that they eat of them (for they are not held wholesome; although the Spaniards, as I have seene, doe eat them), but to recreate themselves, and in revenge of the injuries receiveth by them; for they live long, and suffer much after they bee taken, before they dye.

'At the tayle of one they tyed a great logge of wood, at another an empty batizia [small cask], well stopped; one they yoaked like a hogge; from another they plucked out his eyes, and so threw them into the sea. In catching two together, they bound them tayle to tayle and so set them swimming...'

Sailors 'punish' a shark for the misdeeds of its species in this 1840 engraving.

THE FINAL PARADE

The story of the wreck of the Birkenhead is notorious in the annals of shark attacks. No one can say for certain how many of the 455 who died were taken by sharks.

At 2 am on 26 February 1852 the paddle frigate *Birkenhead,* carrying 490 soldiers of the British Army, with 25 of their wives and 31 children, together with a crew of 134, struck a reef about 1.6 km (1 mile) off Danger Point near the southern tip of Africa.

In the first few minutes following the collision, confusion reigned. Seamen, soldiers and passengers struggled onto the deck to escape water rapidly filling the ship as she settled at the bow.

On deck the officers moved quickly to establish calm, ordering the men to fall in on the poop. Lieutenant Colonel Alexander Seton, the senior army commander aboard, called his officers together and calmly requested them to make sure that any orders given by the ship's captain – Robert Salmond – were instantly obeyed.

Captain Salmond ordered the women and children into a lifeboat, detailing an ensign and a sergeant to forcibly separate women who

clung desperately to their men. A second lifeboat with 30 men aboard was lowered into the swell. For the 600 remaining aboard the dangerously listing ship there were no lifeboats. Many trapped below decks had already drowned. Many more had been killed by falling wreckage or swept overboard. From the water came terrified screams as floundering swimmers were dragged down by packs of sharks that cruised around the doomed ship. The 200 survivors still able to stand, supporting those who were not, stood fast on the poop deck.

Salmond climbed a few metres up the mizzen rigging and shouted to the men: 'Save yourselves. All those who can swim jump overboard and make for the boats. That is your only hope of salvation.' Lieutenant Colonel Seton, was appalled. The lifeboats were already dangerously

full. If 200 soldiers tried to board them, the boats would be lost.

Seton raised his hand above his head and shouted: 'You will swamp the cutter containing the women and children. I implore you not to do this thing. I ask you all to remain where you are.' Some survivors later said that three men went over the rail, but of the rest of the 200, not a man moved. They stood rigidly to attention. Moments after Seton's command, the *Birkenhead* broke its back, the bow slid beneath the water and the stern reared up. A surviving officer wrote later: 'Every man did as he was directed and there was not a cry or a murmur among them until the vessel made her final plunge... [The officers] had received their orders and had carried them out as if the men were embarking instead of going to the bottom of the sea.

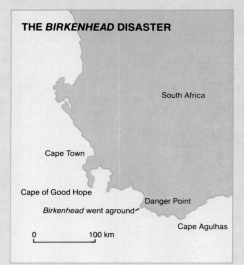

THE *BIRKENHEAD* DISASTER

South Africa

Cape Town

Cape of Good Hope

Birkenhead went aground

Danger Point

Cape Agulhas

0   100 km

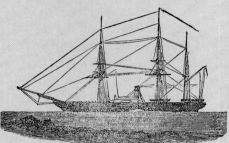

THE "BIRKENHEAD" NEARING THE SUNKEN ROCK.

STRIKING THE ROCK.

BREAKING.

THE WRECK.

There was only one difference – I never saw any embarkation conducted with so little noise or confusion.'

The *Birkenhead* went down just 30 minutes after striking the rocks. Within a short time all but a handful of those that had survived the sinking were dead. On the surface of the water, stained crimson with blood, floated the barely-recognisable remains of those torn apart by the sharks.

Lieutenant Frank Girardot later described what happened in a letter to his father. 'I remained on the wreck until she went down. The suction took me down some way and a man got hold of my leg, but I managed to kick him off and came up and struck for some pieces of wood... I was in the water about five hours... the surf ran so high that a great many were lost trying to land. Nearly all those that took to the water without their clothes on were taken by sharks; hundreds of them were all around us and I saw men taken quite close to me, but as I was dressed... they preferred the others.'

More than 60 men made the 1.6-km (one-mile) swim to shore and safety – though most of the men aboard the *Birkenhead* could not swim, including Alexander Seton who drowned. Captain Salmond was thrown overboard and died in the water when struck by a piece of falling wreckage. The wreck claimed 455 lives, but the proportion claimed by the ship, the sea and the sharks is one of the mysteries of that dreadful night. One of the survivors, Captain Wright, said later that, but for the discipline of the men, the loss of life would have been still higher.

Four contemporary engravings show the sinking of the *Birkenhead.* Fifty survivors clung to the rigging on the mainmast after the ship sank. When the schooner *Lioness* reached the scene at dawn on the following day only 30 were left, the rest had died from cold and exhaustion. News of the disaster reached London in the second week of April 1852, and these engravings appeared in the *Illustrated London News* on the 10th. The same issue carried a melancholy list of those who had perished, but only passing mention was made of the sharks.

WHERE SHARKS ARE GODS

Sharks play an important part in the lives of many of the island peoples of the Pacific Ocean. Shark gods and goddesses are still worshipped, although human sacrifices are no longer made to them. The strange practice of shark calling was also widespread, and still takes place today. Here author Olaf Ruhen describes his experiences in the Pacific.

Numerous magic rituals accompany any attempt at shark calling. Gifts must be made to the appropriate gods, and the fish-magic man will be called upon to bless the dangerous enterprise.

In the reign of Edward II, King of England from 1307 to 1327, and throughout the following centuries the sturgeon, comparatively rare in English waters, was proclaimed a Royal Fish: it belonged to the King when captured. In the waters surrounding the Trobriand Islands, in the Solomon Sea, a similar distinction was accorded the shark. Any ambitious lad, eager for acclaim or privilege, would set out to catch a major shark and deliver it to the king at his village of Omarakana on the island of Kiriwina. If he did this successfully he certainly proved himself a man to watch. It was the crowning achievement of his adolescence.

It was by no means an easy accomplishment, for tradition ruled that the catcher had to follow certain routines and that he had to be alone in his canoe. About 1954, my wife and I were staying with the almost legendary island trader Mrs

Amy Lumley in her house at Gusoweta on Kiriwina's southern coast, when village gossip brought the news that a lad named Benia had the enterprise under way, and I went to see his preparations, which centred on a small one-man outrigger canoe. Besides the regular equipment of paddle and bailer shell it held a cane hoop one metre (3.3 ft) in diameter, on which were threaded nearly 20 of the hard shells of coconuts, in random order, but moving freely on the cane.

There was also something that at first glance gave the impression of a two-bladed aircraft propellor, carved from a large branch of the native hibiscus, a tough pale wood as light as balsa, and rather more than two metres (6.6 ft) long. The paddle ends were not designed to work in any reciprocal or complementary rhythm. The thickening at what would have been the boss had been pierced and threaded with

about four metres (13 ft) of rope laid up from pandanus-root fibre.

The fish-magic man had taken time from his regular employment in the mission garden and was in process of blessing the enterprise, chewing up the words he addressed to the gods and ancestors with a piece of ginger root that lay in a cup of calophyllum leaves he pressed close to his lips, and expelling the abundant saliva that resulted in a fine spray directed at the canoe and its contents. This, without explanation, was all I was permitted to see at this time, but I noticed the gifts of betel nut and cooked yam that the lad had made to the gods.

Rather shaky deduction assisted by luck afforded me a view of the more active stages that came later, not a good view, for I watched from the top of a low coral cliff.

Benia had the cane hoop half submerged in the water and was holding it by the top to shake it

The shark caller's equipment. The rattle is made from a cane hoop, about one metre (39 in) in diameter, threaded with some 20 discs cut from the hard inner shells of coconuts. Held half submerged in the water, it is shaken vigorously so that the noise will attract a shark's attention.

A shark's eye view of the source of the strange vibrations that have attracted its attention. The fish lure invites the shark to puts its head through the rope noose. To grapple, bare-handed, with an immensely powerful, three-metre (9.8-ft) shark from such a small craft is obviously highly dangerous, and sometimes leads to the death of the would-be hero.

vigorously. After only a few minutes a stiffening of his body announced that he had seen the target, and a moment later I saw it too: a good sized shark, its dorsal fin occasionally breaking the surface as it changed direction, half-right, then half-left but closing in on the canoe with some speed. Along the side of the canoe, half-obscured by Benia's body, the propellor shape was secured by a fragile lashing to the protruding tops of the ribs. When the shark was committed to its final charge Benia pulled the hoop rattle inboard and tossed it over the lashings of his outrigger, out of the way. He had made a slipknot of most of the rope attached to the float and now it opened in a big circle in the water beneath him, and he took up his lure, a fish fastened to a short pole.

The shark, now near the canoe, continued what must have been its questing investigation into the cause of the surface disturbance and underwater sounds that had brought it up, and Benia thrust his fish-on-a-stick through the loop, keeping that open with a light pressure on the slipknot.

I would have given much for a closer look at the action of the next minute or so – it took most of Benia's ingenuity and a share of his patience to inveigle the shark, but eventually the big fish thrust through the loop, Benia pulled it tight over the slatted gills and then the action exploded.

The float took erratic life. It twisted back and forth against the panicky struggling of the shark, ever coercing it to the surface. Benia's task was now nothing more than to keep the canoe free of the turmoil and watch his prey as it tired. An hour passed like a minute before the lessening spasms of frantic fury subsided altogether and the shark remained under the surface. Benia paddled carefully up to the float where it now lay on the surface, twitching only occasionally. He exchanged the paddle for a shark club and, pulling the head out of the water, though with a great deal of caution, despatched the great beast with a few well-aimed blows. My imagination had to supply the yells of victory – I could see the action but the sound did not carry to the cliff. The shark was a good one, a grey nurse, three metres (9.8 ft) long, longer anyway than the canoe from which Benia, un-aided, had encompassed its defeat.

This capture of sharks after calling them in occurs in several other locations: the St Georges Channel, for example, that separates New Britain and New Ireland. It has been reported from Kiribati and Tonga, and certainly there will be many communities from which the bold and vigorous youths take chances against the

Coir armour worn by Gilbert Islanders while fighting with shark-tooth swords (Kiribati).

Sword studded with shark teeth (Kiribati).

Sword studded with shark teeth (Kiribati).

Coir helmet worn by Gilbert Islanders (Kiribati).

frequent ferocity of certain sharks, notably the wide ranging tiger.

In other island groups the shark was actually worshipped. In Savo Island in the Solomon group he was regarded as a beneficent deity and in the northern districts of Malaita, as well as nearby in Gizo, and in the Roviana Lagoon to the north there are many tales of sharks having come to the rescue of fishermen who had had ill-luck with their craft. In the coral-enclosed Langalanga and Lau lagoons on Malaita the shark is revered too, although the main gods are the ancestors. These people share with the Tongans and some others the comfortable belief that sharks will not eat them. Though Benia's shark, like the English sturgeon, came rightfully to the King's table, respect for the divine element in the shark's body was sometimes so extreme as to forbid the eating of the flesh.

In the Fijian theogony, chief of the gods was Degei, who was also an ancestor and historically a leader of the tribes that first settled the islands. He is known as 'the origin of the people' and he is envisaged as a huge snake who lives in a cave in the northernmost peak of the Kauvadra Range in the north of Viti Levu. The Reverend J. Water-house, writing about 1860, noted that by extension all snakes were known and honoured as 'the off-spring of the origin'. But Degei's existence was nothing more than an orgy of sleeping and eating, and he took no interest in human affairs.

The executive gods were seen as Degei's sons and were recognised in different animals or different features of the landscape, command-ing some reverence. But none exercised a greater puissance than Dekuwaqa, who was seen in a large shark rumoured to live in a sea-cave under the temple erected on Benan Island.

In Taveuni, Fiji's 'Garden

Island', he was recognised in a bask-ing shark 13 m (42.6 ft) or more in length that used the rocks at the north end of the island as scratch-ing posts to rid itself of barnacles. The people of Natewa and Cakau-drove were concerned to keep on friendly terms with him. In the old days he piloted them on night raids, the canoes following the phosphor-escence of his wake, and for this reason he was also called Daucina 'the giver of light', the name by which he was known at Levuka and Kadavu. He was the chief god of fishing and seafaring com-munities, and from his love of fair women he was also noted as the god of adulterers.

In his honour the eating of

A group of objects, collected from various parts of the Pacific, that are either involved in shark worship or incorporate the parts of captured sharks. Some objects, such as swords edged with shark's teeth, can still be purchased today, although they are not as elaborate as the original weapons intended for use in battle.

Coconut shell rattle used to attract sharks (New Britain).

Shark lure (Gulf Province, Papua New Guinea).

Wooden float, with rope attached, to prevent captured shark from diving (New Britain).

Carved shark inlaid with pearl shell (Solomon Islands).

Shark-tooth knife (Kiribati).

Shark-catching charm (East Sepik Province, Papua New Guinea).

Lau) on the death of the previous incumbent, Ratu Pope. Two days before that event he was at sea and he made what might have seemed an innocuous entry in his diary:

'Soon after 11 am a specie of hammerhead shark (seven foot) [2.1 m] seen astern. He stayed over an hour alongside, six or seven times allowing himself to be touched, patted and pulled.'

Dr S.M. Lambert, who wrote the book *A Doctor in Paradise*, recorded that Ratu Sukuna told him that he knew at once that Ratu Pope was dead, or would soon die.

In his biography of Ratu Sukuna, Deryck Scarr commented: 'I joked about synthetic pillars of the Church who still believed in the heathen temples. Nonetheless the enormous satisfaction he drew from administering the land of Fiji could hardly have been separable from at least a sympathetic feeling for the spirits of the land. He was culturally enriched by remembering the places where spirits dwelt as also by the legends of Dekuwaqa.'

It seems doubtful whether tributes to Dekuwaqa amounted to a true worship or anything much more than a recognition of the commmon ground shared by all life – animal or vegetable. The latter is represented by the sacred groves of Ivi trees *Inocarpus edulis*, the Tahitian chestnut, or the ironwood which is very often grown in plantations around graves. Or perhaps these things are simply reminders of such spiritual power as rules humans too.

Fear is a motive for establishing a god, and man has worshipped at one time or another most of the living things he fears. History gives few examples of the deification of a fish. In the middle-east a notable exception was Dagon, the national god of the Philistines, a fish with the hands and face of a man for whom there were temples built at Gaza and Ashded, and who later

sharks was tabooed. Sometimes in the tidal fish-traps a shark was taken with markings on its belly claimed to be the tattoos of Dekuwaqa. Such a fish was freed into the open sea. When Fijian vessels passed over areas of sea that he was known to frequent the crews poured libations into the water, and accompanied this with offerings of cooked food. Remnants of this custom still persist.

The wholesale conversion of the Fijians to Christianity did not eliminate such deference to the old gods. Ratu Sir Lala Sukuna, one of the most distinguished Fijians of the twentieth century, an avowed Christian with an Oxford education, became Tu'i Nayau (the Lord of the gained a following in Phoenicia and parts of South Palestine. But Dekuwaqa seems his only rival in the records of history. Perhaps he gained a hold because of the scant population dependent on his share of ocean, and the few other living threats those people found to menace them there.

It is certain that the introduction of other religions tended to drive the shark gods from their sphere of influence. They had considerable power less than a score of years ago, particularly in the heart of the Solomon Islands. At that time shark gods commanded a majority of the Savo Islanders, who lived on the flanks of a graceful volcanic cone about halfway between Guadalcanal and Tulaghi in the middle of New Georgia Sound. Here I heard tales in abundance of this kindly swimming god who rescued the islanders from shipwreck, towed them home as they clung to its dorsal fin, rescued children who fell overboard unnoticed late at night – and was selective enough to destroy its worshippers' enemies.

Shark worship has become much less popular on Savo, a circumstance attributed to the number of corpses that had come the way of the local sharks during naval battles in World War II. It still persisted on Tulaghi, on the northern parts of Malaita and apparently in many other remote places. Some women with whom I travelled from Bellona Island in the far west of the Solomons had dark tattoos between their breasts, and in these sharks shared honours with frigate birds. Bellona gods had no competition at all from the missionaries until the 1940s, but I was told that shark worship was hale and hearty in the village of Laulasi in the Langalanga Lagoon.

This was actually two villages, crowded onto artificial islands built on the reef or in the shallow waters of the lagoon by the industrious hands of their inhabitants. One island was Christian, one heathen, but both were alike in their hordes of laughing children, most of whom could swim and handle a small canoe before they could walk.

In the heathen village the women all faced the one way, and were very pernicketty about it too, walking backwards or sideways if they wanted to go anywhere else. Not one faced the northern end of the village where, in carefully stencilled letters, a prominent notice proclaimed: PERSONS IN BLACK CLOTHING NOT PERMITTED IN THIS AREA. The

A tropical dusk descends over the island of Laulasi in the Solomon Islands. The inhabitants of this artificial island, built on a submerged reef, keep alive a tradition of shark worship that has a long history.

forbidden tract contained three large communal houses, each with a notice painted on its gable. They read: HEADMASTER RU GOLA, HEADMASTER MAEMADAMA, HEADMASTER MOUSI. These named individuals had nothing to do with schools, but were pagan priests controlling the 'spirit houses'. Maemadama, in a short wraparound skirt and wearing a flower in his hair, gave me permission to enter the one marked with his name.

The narrow doorway barely admitted enough light for vision. White wood ash around a central fireplace testified that a large congregation had gathered the night before. When I moved towards a big collection of leaf-wrapped shells tidily disposed in niches, Maemadama restrained me.

Indeed I was glad enough to return to the sunny exterior, where hordes of children bobbed about the lagoon surface in tiny canoes with only a few centimetres of freeboard, and gossiping women sat over their chores in the narrow passageways between the huts in their bright cladding of pandanus leaf. I had noted that although the overt worship was directed to the ancestors, these, in some mysterious way, also represented the sharks.

Now that there is no longer any mystique attaching to factory-made goods, the headway of the imported religions too is diminishing. The Christian community has enjoyed advantages from its conversion – health, education, and a boat-building school which concentrates on the techniques of repair – but such benefits have also come the way of the heathen by a process of osmosis. And perhaps the heathen have had to do more thinking for themselves, without missionaries to provide instruction.

Wade Doak, in his book *Sharks and other Ancestors,* reports on a visit to Laulasi a decade after mine. The shark worshippers have formed a company 'Laulasi Adventure Tours', properly constituted with seven directors and fifty shareholders in the village. They charge the tourist $19 a day to visit, and they provide transportation from the Malaita capital of Auki in outboard-propelled canoes. And they give good value for the money. It seems the shark gods have no argument with modern ways.

Through the medium of these skulls – the remains of long-dead ancestors – shark priests on Laulasi call up the spirit sharks. Few, if any, Europeans have been allowed to witness the rituals associated with shark worship.

THE SHARK PAPERS

*The strange story of the capture of the **Nancy**, and the undoing of her captain, is a testament to the extraordinary appetites of sharks. This is the first of two occasions (that concerning the Shark Arm Case is told on p 114) in which sharks have been instrumental in bringing human villains to justice.*

The *Nancy,* a brig owned by German-American Maryland traders, left Baltimore on 3 July 1799 bound for Curaçao and Haiti with a cargo of German dry-goods, food and timber. She was commanded by Thomas Briggs, a resourceful sailor, and one of the unluckiest in the history of the United States merchant marine.

Briggs delivered his goods to Curaçao and was on his way to collect the return cargo of coffee at Port-au-Prince, Haiti, when on 28 August the *Nancy* encountered the British cutter, *HMS Sparrow,* part of a blockade of Haiti effected by the British Navy. The United States and Britain were at war at the time.

Briggs immediately crowded sail onto the two-masted, square-rigged *Nancy* and fled. The *Sparrow* gave chase, put a shot across the *Nancy's* bows and boarded her – but not before Briggs had dumped the ship's papers overboard and replaced them with Curaçao-made forgeries stating that the *Nancy* was Dutch-owned, and therefore free to pass through the blockade.

The *Sparrow's* Commander, Lieutenant Hugh Wylie, was not impressed by the forgeries, or by Briggs' threat of legal action. He told Captain Briggs that he, the *Nancy* and her crew were now a prize of war. He ordered his men to sail the *Nancy* to Port Royal, Jamaica, where Briggs could put his case that the seizure was unlawful to the Vice-Admiralty Court.

A suit for salvage was brought on 9 September against 'a certain Brig or Vessel called the *Nancy,* her guns, tackle, furniture, ammunition and apparel, and the goods, wares, merchandize, specie, and effects on board her, taken and seized as the

The jaws of the shark which swallowed the *Nancy's* papers were preserved, and for a time were on display in a London museum. The shark involved was probably a tiger – a species renowned for its remarkable appetite (see p 117).

property of some person or persons being enemies of our Sovereign Lord the King ...' On 14 September, Briggs put his case for dismissal of the suit with costs, backing it with an affidavit in which he swore that 'no papers whatever were burnt, torn, thrown overboard, cancelled, concealed, or attempted so to be ... all the papers on board [the *Nancy*] were entirely true and fair.'

A bizarre event at sea 15 days previously – the day after the seizure of the *Nancy* – was about to undo Briggs. On that day, another British ship, *HMS Ferret,* encountered a dead bullock in the sea near San Domingo Island. It was surrounded by sharks tearing at the carcass. The *Ferret's* commander, Lieutenant Michael Fitton, had a taste for shark fishing, and hooked the biggest of the sharks. When they decked and opened it, Fitton's men discovered inside the shark the wholly preserved ship's papers of the *Nancy,* tied in string. Fitton dried them in the sun on deck. By coincidence, Lieutenant Wylie, the *Sparrow's* Captain, had been invited for breakfast aboard the *Ferret* that very morning. The astounded Wylie, together with Fitton, immediately

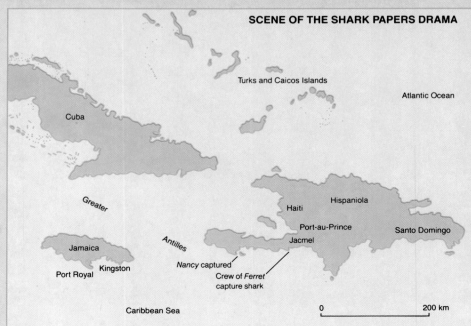

SCENE OF THE SHARK PAPERS DRAMA

Turks and Caicos Islands

Atlantic Ocean

Cuba

Greater

Jamaica

Port Royal   Kingston

Antilles

Nancy captured

Crew of *Ferret* capture shark

Haiti

Hispaniola

Port-au-Prince

Jacmel

Santo Domingo

Caribbean Sea

0   200 km

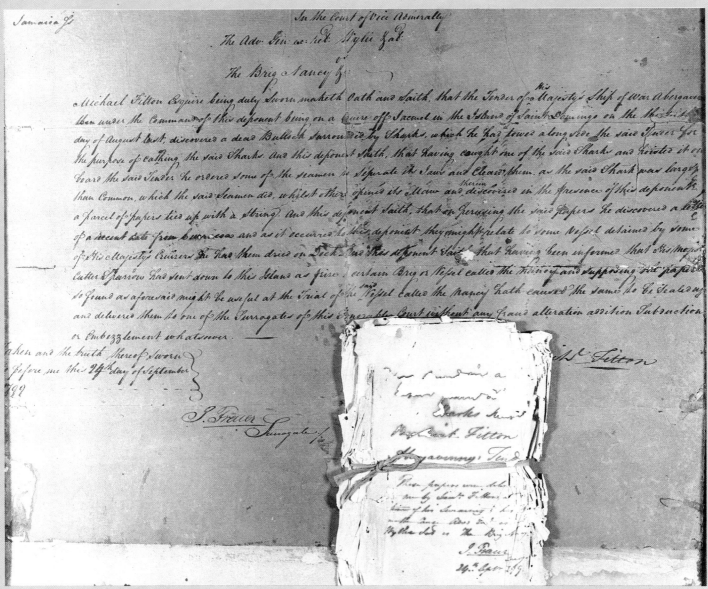

The *Nancy's* original papers, bound in string, and Lieutenant Fitton's affidavit which detailed the strange way in which he obtained them.

turned for Port Royal. On 24 September, Fitton's affidavit, giving his account of the shark's capture and the discovery of the papers and his supposition that 'the papers so found as aforesaid might be useful at the trial of the said vessel called the *Nancy* hath caused the same to be sealed up and delivered' was sworn in the Vice-Admiralty Court. The resourceful Briggs' perjury was exposed. The *Nancy* became a lawful prize of war. British justice had prevailed over an American ship running a blockade of Haiti in

a court sitting in Jamaica – with the help of a freakish stroke of luck and a shark.

The *Nancy's* papers are exhibited in the Institute of Jamaica at Kingston. The shark's jaws were, for a time, set up ashore in Jamaica where they carried the inscription: 'Lieutenant Fitton recommended these jaws as a collar for neutrals to swear through'. Until recently they were held by a small museum in London, but the collection has now been dispersed, and the jaws have disappeared.

The *Nancy's* original papers, bound in string, and Lieutenant Fitton's affidavit which detailed the strange way in which he obtained them. Among the papers are a number of letters written in German from the ship's owners instructing their agent on board, Christopher Schültze, on what to do in certain circumstances. The compromising nature of these letters – which proved that the ship was American property, and that there was a conspiracy to conceal her identity – made it essential that they did not fall into British hands. Wrapped around the papers is a note signed by John Fraser, an official at the court in 1799, certifying their authenticity. These documents are now preserved in Kingston.

SHARK ARM MYSTERY

The Shark Arm Murder is still one of the great unsolved riddles of Australian criminal history. It was only through the fateful intervention of two hungry sharks, and an extraordinary series of coincidences, that the police even knew a serious crime had been committed.

On 25 April 1939, a 4-m (13-ft) tiger shark was listlessly cruising in the Coogee Aquarium, Sydney, under the fascinated gaze of 14 holidaymakers. Suddenly the shark, which had been tired and languid for three days, lashed its tail, bumped hard against the white-tiled pool wall, swam towards the shallow end, turned in three irregular circles and vomited up a bird, a rat – and a muscular human arm. A cloud of evil-smelling brown fluid spread over the pool floor.

When he answered the telephone at the Randwick Police Station a few minutes later, Detective Constable Frank Head heard the outline of an interesting, but fairly routine, mystery. The central questions – whose arm was it, and how had it found its way into a tiger shark's stomach? – might be quickly answered. It might be the arm of a suicide who had ended his misery in the harbour. It might be part of the torn remains of a derelict who had stumbled off a rocky cliff.

At the aquarium, Head got his first intimation that the case might not be routine: a rope about 600 mm

A "Truth" in Every Home: Nett Sales Per Issue Largest in Southern Hemisphere

Truth

FORECAST: Mostly fine. Cold night. Variable winds.

No. 2365.

Registered at the General Post Office, Sydney, for transmission by post as a Newspaper.

SYDNEY, SUNDAY, MAY 5,

THE PUBLIC LIBRARY
7 MAY 1935
OF NEW SOUTH WALES

WHAT SICK SHARK REV

MR. BERT HOBSON and his son Ron, who captured the shark. Above is the monster, and on the right the shark is seen in the Coogee Aquarium Baths.

Murder

BY one chance in a million — the fact that little more than a week ago a sick shark in captivity at Coogee Aquarium disgorged a tattooed human arm — some of the best brains of the Sydney C.I.B. are convinced that Fate has revealed a cold-blooded murder, which like the case of the Albury Pyjama Girl, may prove to be another perfect crime. The victim of this mysterious death drama is James Smith, of Gladesville, former billiards marker at City Tattersall's Club, a well-known suburban billiards saloon keeper, one-time promising lightweight boxer, and a man with seemingly not an enemy in the world.

"Truth," which played an outstanding part in the identification of Smith, knows that the dead man's relatives have expressed grave fears as to how he came to meet such a shocking fate.

They scoff at all theories of suicide. Instead, openly they have expressed the opinion that Jimmy Smith was murdered.

The finding of the arm opens up theories as to how he could have met such a fate.

Smith was either the victim of a quarrel while on a fishing excursion, and after being clubbed into unconsciousness, was securely bound and tossed into the ocean's depths, or—the dastardly deed was committed on dry land, and the brain of the human monster responsible conceived the idea of covering up traces of the crime by sending him to a watery grave.

TATTOO MARK.

The police were impressed and there was much to make them interested.

After the arm was disgorged by the giant tiger shark which for a week had lazed about in the clear waters of Coogee Aquarium, detectives discovered a vital clue. The limb bore a crude but discernible tattoo mark, and a knotted rope still hung from the wrist.

The tattoo mark was a distinctive one—that of two boxers sparring up to each other. It was so distinctive indeed that, when the description appeared in last week's "Truth", Jimmy Smith's brother went along at once and informed Detective Head that he was certain that the arm was that of his former boxer brother.

Until that moment, the police (not unnaturally) had believed that the shark's victim was a suicide, someone who, with nothing to live for, had deliberately drowned himself, and had then been swallowed by the tiger of the seas.

But Jimmy's brother soon enlightened them on that score. Jimmy was no coward to take an easy way out . . . and what was much more to the point.

THE TATTOOED BOXERS of Smith's arm.

Jimmy Smith had everything to live for.

Then the police began to make inquiries—inquiries that have been carried out with amazing secrecy, so much so that all the dead man's relatives have been ordered to say nothing at all about the matter. Just how much the authorities regard the crime is borne out by the fact that Detective-Inspector Quinn, Detective-Sergeant Matthews, Detective-Sergeant Allmond (who with Detective-Sergeant McRae) has been for months working on the Albury culvert horror, and Detective McDermott have been assigned to the case.

Every single belonging of the dead man has been temporarily taken by the police in the hope that it will yield some clue to the mystery . . . a mystery made all the more remarkable by the fact that five days after he had left home for the last time, Jimmy

Smith's broken-hearted widow received a telephone message.

That message was delivered by a voice she had never heard before. "Don't worry," said the mysterious caller . . . "Jimmy will be home in three days' time."

Then the caller rang off . . . Mrs. Smith had not the slightest idea then that anything was seriously wrong. Only when the shark was netted, and after being put into the Aquarium (instead of, as it might well have been, killed as soon as it had been dragged ashore) gave up its grisly clue, later to be identified by the dead man's brother, did she realise that the mystery caller must have known full well that Jimmy Smith would never come home again.

Behind their veil of secrecy, the police are hunting . . . hunting . . . but "Truth" to-day is able to tell the exclusive story of the facts that led up to this astounding tragedy—a tragedy such as Edgar Allan Poe never dreamed of in his weirdest fiction . . . a tragedy that is without parallel in the history of crime the world over.

A little more than a fortnight ago, Mr. Bert Hobson, genial proprietor of the popular Coogee Aquarium Baths, set out with his boy Ron in their small boat, coming to anchor about a mile off Coogee Beach.

An expert at landing sharks, Mr. Hobson set one of his big lines. He didn't have to wait long before he got a bite. In no hurry to land the shark, the fisherman played him, but it soon dawned on him that something was amiss.

Away went more line so Mr. Hobson decided to haul in. He has rarely had such a hectic battle and he soon learned the reason. On getting his catch to the top he found that the shark he had hooked had been eaten practically away and tangled, but securely held, in the line was the hungry—held, in the line was the hungry—tacker—a 15 foot tiger shark, fearless and savage member of the species.

The million to one catch, Mr. Hobson described his feat.

(Continued on Page 24)

(Continued on Page 24)

The lurid front page of the Sydney *Truth* gave a full description of the arm, together with a drawing of the tattoo found on it. On the following day the victim's brother, Edward Smith, came forward to identify it.

A sinister shape cruises around the perimeter of the pool at Coogee aquarium. Eight days after its capture this 4-m (13-ft) tiger shark vomited up a human arm with a length of rope attached to it.

Bert Hobson (left), a Sydney fisherman, caught the tiger shark when it became entangled in a fishing line he had left overnight about 1 km (1100 yds) offshore. He towed his prize to the aquarium at Coogee where it was put on display by his brother Charles (right).

(24 in) long was tied around the wrist. The knot, he noted carefully, was a clove-hitch. But at least identifying the person from whom the arm had been severed should not be too difficult: on the forearm was a distinctive tattoo of two boxers in red trunks shaping up.

Nor did Head have any difficulty discovering when and where the shark was caught: Albert Hobson, whose brother Charles owned the Coogee Aquarium, explained that eight days previously on 17 April, he had been shark fishing about 1 km (1100 yds) off Coogee Beach, and had left a baited set line overnight. When he returned next morning he discovered he had caught not one shark but two: a small shark had taken the bait and hook, and the 4-m (13-ft) tiger, in the cannibalistic way of the species, had eaten its unfortunate brother – and then itself become tangled in the line. Albert towed it to the aquarium.

The answer to the next question – had the shark bitten the arm from a suicide or from the body of a drowned man, or from a live swimmer? – dramatically deepened the mystery. Dr Arthur Palmer, the Government Medical Officer for the City of Sydney, examined the arm the day after the shark disgorged it and said it had been severed at the shoulder by a sharp instrument, probably a knife. But it was not the work of a surgeon – there were none of the skin flaps a surgeon would have left after an amputation. Although it was remarkable what terrible wounds lunatics could inflict on themselves, he thought it 'extremely unlikely' that the man could have cut off his own arm. Dr Palmer said he 'could not deny' the possibility that the man from whom the arm had been severed was still alive, but 'I could hardly conceive that he would be'.

The *Truth* newspaper, which specialised in lurid coverage of bizarre crimes, published a photograph and description of the tattooed arm at the end of April. An

The victim, James Smith, whose body was never found. His partially dismembered corpse had apparently been dumped into the sea somewhere off the coast of Sydney.

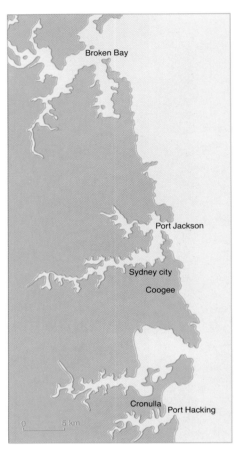

Sydney's convoluted coastline was the scene of the shark arm drama. The aquarium which housed the shark was at Coogee, an outer suburban beach, and Brady's cottage was at Cronulla, a popular resort on the southern outskirts of the city.

English-Australian, Edward Smith, went to the Randwick police station and said he recognised the tattoo and that the arm was that of his brother, James Smith. The crucial tattoo evidence had survived by the million-to-one chance of the small shark being hooked before it digested the evidence, and then the 4-m (13-ft) tiger shark which ate it becoming entangled in Albert Hobson's line.

The Sydney police quickly confirmed Edward Smith's identification of the arm: James Smith's fingerprints were on the police files from his arrest four years previously on a charge of illegal starting-price betting.

Smith, a tall, well-built Englishman who arrived in Australia when he was 20, had had a varied career in Sydney, trying his hand as a bartender, billiard marker, amateur boxer, sports club manager, bookmaker, builder and boat caretaker. Smith had left his Gladesville home on 7 April, 18 days before his arm floated into view at the aquarium pool. He told his wife Gladys that he was going fishing at Port Hacking, on the southern outskirts of Sydney, for a few days with a friend called Greg Vaughan and 'a rich man from interstate'. Smith did not go to Port Hacking. He visited Vaughan, then went to a cottage at Cronulla where

he visited another friend, Patrick Brady, a forger.

Smith and Brady met the year before the 'shark arm' appeared, when Smith was employed as caretaker of the luxury yacht *Pathfinder*, owned by Reginald Holmes, a wealthy boatbuilder and engineer. When Holmes bought the yacht it was insured for $4000. He increased the insured value to $7000 and then connived with Greg Vaughan, who became the nominal owner, to bump it up to $17 000. Brady visited Smith aboard the *Pathfinder* in Broken Bay. The broke former builder and the forger found they had much to talk about, and while the waves of Broken Bay

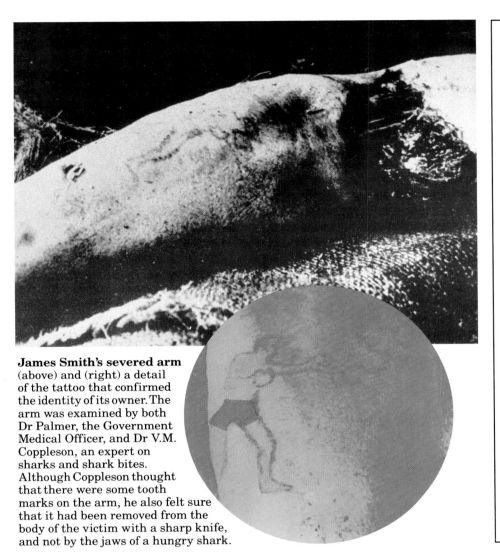

James Smith's severed arm (above) and (right) a detail of the tattoo that confirmed the identity of its owner. The arm was examined by both Dr Palmer, the Government Medical Officer, and Dr V.M. Coppleson, an expert on sharks and shark bites. Although Coppleson thought that there were some tooth marks on the arm, he also felt sure that it had been removed from the body of the victim with a sharp knife, and not by the jaws of a hungry shark.

A distinctively-patterned tiger shark.

THE CURIOUS EATING HABITS OF TIGER SHARKS

The tiger shark is one of three species most frequently named as being responsible for attacks on humans. Voracious eaters, they will swallow almost anything they encounter in the sea. At various times the stomachs of captured specimens have been found to contain an astonishing variety of objects including: a coil of copper wire, nuts, bolts, lumps of coal, boat cushions, clothing, a tom-tom, an unopened can of salmon, driftwood, birds, other sharks, seals and the head of a crocodile.

Sharks can regurgitate the contents of their stomachs at will, and some can apparently store food undigested. Sir Edward Hallstom, honorary director of Sydney's Taronga Park Zoo, once observed this phenomenon in a tiger shark that lived for a month at the zoo in 1950. On two occasions during its captivity the shark was fed on horse meat which it regurgitated. After it died the shark's stomach was cut open and was found to contain two undigested dolphins, eaten before its capture.

gently lapped the *Pathfinder's* sleek hull, their conversation often turned to money, and to schemes which would make them rich men.

They planned to go to Hobart in Tasmania together. Brady went first, looking for buyers for Reginald Holmes' speedboats – but their joint Tasmanian ventures were aborted when Brady forged a cheque for $1000, was arrested and charged – and jumped bail. He returned to Sydney, where he rented the Cronulla cottage. Years later, Brady told a Sydney crime reporter, Vince Kelly, that he and another man had loaded the *Pathfinder* with rocks. On 9 April 1934 a coal ship saw the *Pathfinder* sinking in the

sea off Broken Bay and a man rowing towards the shore. When Vaughan reported the loss to the insurance company, the company called the Criminal Investigation Bureau – and Vaughan dropped the claim. The police suspected Smith's death involved some falling out among the conspirators who had attempted to defraud the insurance company, but could prove nothing.

Gladys Smith was used to her husband's unconventional lifestyle, but by 20 April, when he had been gone nearly two weeks, she was getting worried. She called one of his hangouts and was told James had been there a few days earlier. Four days later, with still no word,

she called Greg Vaughan. Vaughan told her that James had said he was taking a boat to Cronulla.

The police focused on Brady: he was a man with form, a criminal record, and Smith had been with him in the Cronulla cottage at about the time he met the violent death which delivered his left arm to the maw of a tiger shark. On 3 May, Gladys Smith received a letter in James' handwriting, addressed to their son Raymond. 'Keep your mother quiet,' it said. 'I am in trouble. But everything will be all right. Tell your mum I'll have plenty of money soon. They want me. Something in town. Never mind, be a man for me.' It was

signed 'Your loving father Jim Smith', and added: 'Destroy this'. The police believed Brady, a gifted forger, had written the note.

When the police questioned Brady, he showed the style detectives expect from guilty suspects: he lied. He had not seen Jim Smith since he returned from Tasmania, he said. He had not been living at the Cronulla cottage.

The police theory was that Brady had murdered Smith, tried to fit the body into a tin trunk, discovered that it would not fit, then severed Smith's left arm, tied it to the trunk and dumped the trunk at sea. There a hungry small shark had taken the arm with the rope, and an even hungrier tiger shark had intervened, like fate, and later coughed up the evidence at Coogee.

The difficulty that the police had, was that although they had an arm in a condition suggesting that murder was the only explanation, they did not have a corpse. A dead body is usually the starting point for a murder investigation. When the City Coroner began his inquiry into the cause of James Smith's death, Brady's barrister, Clive Evatt, insisted that the law required a body before the inquiry could proceed. The point went to the Supreme Court, where Evatt ridiculed the idea of having an inquest into an arm. What if one arm surfaced at Newcastle, 171 km (106 miles) north of Sydney, and another in Sydney? he asked. Would two different coroners have two different inquests, with perhaps two different verdicts, on the same corpse? Mr Justice Halse Rogers found Evatt convincing: an arm was not enough, he said, there must be a body, the inquest could not go on.

The police had an even more serious difficulty than the lack of a body – the murder of their principal witness, Reginald Holmes.

On 20 May water police on Sydney Harbour spotted a powerful speedboat careering erratically near Pinchgut Island. When they finally overhauled the speeding boat they discovered Holmes at the wheel, a bullet wound in his forehead, and an empty brandy bottle behind him. 'They say I've been squealing', Holmes gasped, 'but you know I haven't. But I'm going to tell you all about it now.' Holmes was in hospital under police guard for four days while surgeons removed a .32 bullet from his forehead. When he left hospital he went straight to CIB headquarters and made a statement which, if a jury believed it, would send Patrick Brady to the gallows.

Brady had been extorting money from him, Holmes said. Early in April, Brady telephoned

It was at this small rented cottage in Cronulla that James Smith was last seen alive. The police suspected that Patrick Brady had murdered Smith there, and had then dumped his body at sea in a trunk.

Patrick Brady, a forger and small-time crook, was the chief police suspect in the case of Smith's murder. In September 1939 he was tried and acquitted, and he maintained his innocence to the end of his life.

Reginald Holmes, a key witness in the police case against Brady, was murdered on the morning of the Coroner's inquest into James Smith's death. Although two men were later charged, the identity of his murderer, and that of James Smith's, remain unknown to this day.

and said he was coming to see him on an urgent matter. When he arrived, Holmes gave him a brandy, and Brady, dirty, unkempt and distressed, said: 'I had a row with Smith and I have done him in.'

Holmes: 'What are you saying?'

Brady: 'I have killed the bastard. I put his body in a tin trunk and sank the trunk out at sea outside Port Hacking Heads. If you tell the police I've done this, I'll murder you too. And if I can't murder you, one of my mates will. I want you to remember this: if anything comes out about Smith, I know nothing about him. And I don't know you at all.'

Holmes, who had denied knowing Brady in an earlier interview, said he had done so because 'this was what Brady had told me to say; I was terrified about what might happen to me.'

At 1.10 am on 12 June, the day

the Coroner's inquest started, a police constable patrolling an area near Sydney Harbour Bridge noticed a car parked with its headlights on and the near-side front door open. The driver, slumped over the wheel, his hat on the back of his head, was Reginald Holmes. He had been shot in the chest from a distance of 300 mm (12 in) with a .32 calibre revolver.

Patrick Brady was charged with James Smith's murder. When he was committed for trial he stood in court and said 'Jim Smith is one of my best friends...If he is dead I didn't kill him.' At the end of the evidence in the Supreme Court trial, the judge directed the jury to acquit Brady, saying 'if the accused was convicted on the evidence we have heard, the conviction would not be allowed to stand.'

When he spoke to crime reporter Vince Kelly in 1962, Brady

argued that since he was acquitted of murdering Smith, and could never be charged with the murder again, even if he confessed, why should he lie now? 'The doctors tell me I haven't got much longer to live, so I want to tell you all I know about the case. But with my dying breath I repeat: I didn't do it.'

Two men were twice tried for the murder of Reginald Holmes. The first trial resulted in a hung jury; in the second, the jury acquitted both men. Patrick Brady died on 11 August 1965. Vince Kelly went to see his widow the next day and she told him: 'Pat was a weak man, but he wasn't a killer.'

The tiger shark remained sick and lethargic in the white Coogee Aquarium pool for three days after its digestive system revealed the gruesome mystery; then the Hobson brothers killed it. Its stomach contained nothing else of interest.

DEATH IN NEW JERSEY

In 1916, in the middle of World War I, an extraordinary series of shark attacks in the United States of America left three men and one boy dead, with a second boy badly injured, in the space of just 12 days. The sleepy towns on the New Jersey coast, just south of New York, were suddenly the focus of national interest, and the 'shark menace' even claimed the attention of President Wilson and his cabinet.

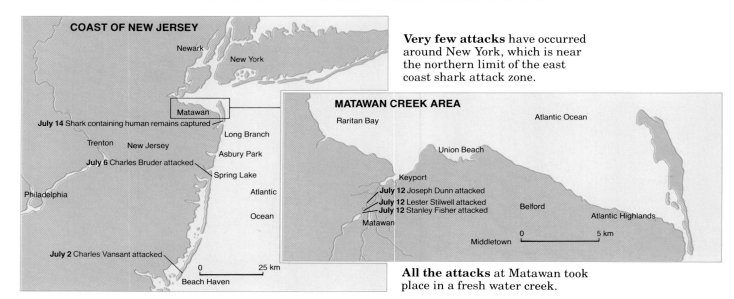

COAST OF NEW JERSEY

Newark
New York
Matawan
July 14 Shark containing human remains captured
Long Branch
Trenton   New Jersey
Asbury Park
July 6 Charles Bruder attacked
Spring Lake
Philadelphia
Atlantic
Ocean
July 2 Charles Vansant attacked
0   25 km
Beach Haven

Very few attacks have occurred around New York, which is near the northern limit of the east coast shark attack zone.

MATAWAN CREEK AREA

Raritan Bay
Atlantic Ocean
Union Beach
Keyport
July 12 Joseph Dunn attacked
July 12 Lester Stilwell attacked
July 12 Stanley Fisher attacked
Belford
Matawan
Atlantic Highlands
0   5 km
Middletown

All the attacks at Matawan took place in a fresh water creek.

The panic came slowly. On 1 July 1916 a 25-year-old Philadelphia fine arts graduate, Charles E. Vansant, was swimming 15 m (16 yds) from the shore at Beach Haven, New Jersey in the United States, when people on the beach saw a black fin slicing towards him and shouted at him to get out. Vansant splashed madly for the beach. As the shark closed on him, he screamed for help, then went under. An Olympic swimmer on the beach dashed into the water and pulled the badly mauled Vansant ashore. He died in hospital the following night.

The tragedy stirred no great unease along the Jersey shore, and still less in Manhattan, where the editors of the *New York Times* placed a two-paragraph report of the attack on page 18. It was one of the minor tragedies of summer, like a swimmer hit by lightning.

Five days later the level of anxiety on the Jersey shore made a

quantum leap upwards. On 6 July, Charles Bruder, a bellboy at a Spring Lake, New Jersey hotel, was attacked while swimming beyond the lifelines. A woman on shore cried that she saw a red canoe beneath the waves, but the 'canoe' was pumping from Bruder's severed legs. He died on the beach. Lifeguards and old Spring Lake residents told a *New York Times* reporter that this was the only shark attack death at the resort that they could remember. The *Times* editors placed the story, with details of women 'fainting' and 'panicking', on page one. Spring Lake hotel staff raised $1000 to send to Bruder's mother in Switzerland.

Although Bruder's death kept swimmers close to shore at Spring Lake the next day, even this elementary precaution extended no more than a few kilometres away. But the businessmen of the local resort towns, naturally assuming that

news of shark attacks would not help to attract summer holiday-makers, wanted reassuring official action – and Spring Lake Mayor, Oliver H. Brown, immediately established a motor boat patrol. The boats dragged bleeding quarters of lamb as bait while marks-men with rifles stood ready to shoot should a fin materialise. None did. The Mayor also ordered the beach bathing area enclosed in a sharkproof wire net, a precaution also taken by the nearby resort of Asbury Park.

Fishermen and surfers ridiculed these precautions, pointing out that no shark had ever been sighted off Asbury Park. Academic experts were also reassuring: Dr A. T. Nichols, the American Museum of Natural History's shark expert, said that there was 'very little chance' of a shark ever attacking anyone. The museum's director, Dr Frederick A. Lucas, said there was more chance of being struck by lightning than being attacked by a shark, and in any case, sharks' jaws were not powerful enough to bite through a human legbone. The captain of a trans-Atlantic liner said he was 'astounded' to learn of a man-eating shark off the New Jersey coast: this was the first time he had heard of one north of the Bahamas. A report in the *New York Times* of 11 July said 'Tiger sharks will hold but little terror for bathers in the waters hereabouts within the next few days' because of the netting at Asbury Park.

It was a classic piece of predictive journalism: within 24 hours sharks were part of every conversation on the Jersey shore.

July 12 brought tropical heat and humidity to Matawan, New Jersey, a small town 17 km (11 miles) west of the Atlantic shore, linked to the ocean by a meandering tidal creek. At 2 pm Lester Stilwell, a 12-year-old boy employed at the town's sawmill, was given the afternoon off because of the stifling heat. With four friends, he headed for the Wyckoff Dock, a dilapidated old steamboat pier on Matawan Creek which was the town's most popular swimming hole – throughout the summer naked boys would liberate themselves from heat and boredom by playing on the exposed dock pilings.

Shortly before Lester and his friends headed for the creek, Captain Thomas Cottrell, a retired sailor, was walking across a new Matawan Creek bridge about 750 m (833 yds) downstream from the swimming hole. He saw beneath the sparkling creek surface a huge black shadow moving quickly upstream with the incoming tide. Cottrell did not stop to tell himself that no shark could be that far upstream, he ran for a telephone and called the town's barber, John Mulsonn, who was also the chief of police. Then he ran to Main Street telling groups of boys headed for the creek,

It was near this spot on Matawan Creek that Captain Cottrell saw the shark. People in the town who scorned the idea that a shark would enter such a narrow waterway, and swim nearly 16 km (10 miles) from the sea, were tragically wrong on this occasion.

Searchers probe the muddy bottom of Matawan Creek in a vain attempt to find the body of 12-year-old Lester Stilwell. The boy's body eventually came to the surface three months later.

merchants and their customers: 'There's a shark in the creek!' People thought he was crazy: a shark? in a creek 10 m (11 yds) across at its widest? Clam diggers worked the shallows at low tide: poor old Tom's eyes must have been playing tricks on him. As Chief Mulsonn stropped his razor he reflected that people tell policemen some crazy stories.

Lester Stilwell, who suffered from 'fits', was a strong swimmer. He was floating further away from the pier than his friends when they saw him suddenly disappear, re-emerge, scream, then disappear in a flurry. His friends sprinted into Matawan shouting that Lester had had a fit in the creek and had disappeared in the water.

Stanley Fisher, a likeable, popular man of 24, who had just started a dry-cleaning business, was one of the many who heard the boys' shouts. Stanley, a 95-kg (210-lb)

blonde giant, had recently been mocked by his friends for accepting a $10 000 life insurance policy instead of cash in payment for a suit. He was nuts, they said. You? Life insurance? At 24? You must be joking. The good-natured Stanley just smiled quietly.

As Fisher ran for the creek, he passed a woman acquaintance, a Matawan teacher, who shouted: 'Remember what Captain Cottrell said. It may have been a shark.' Fisher barely paused. 'A shark here?' he said. 'I don't care – I'm going after that boy.'

Men, women and children were streaming from the town to the pier, among them Lester Stilwell's parents. As Fisher yanked on his bathing trunks and plunged into the creek about 200 people lined the banks while men in rowing boats poled for Stilwell's body.

The urgency of the action, the

need to do something – anything – in the face of tragedy must have driven Stanley Fisher, for in his rational mind he surely knew that if Lester was still beneath the creek he must have drowned, or worse. But the urgency of the moment was very powerful: several other men were also in the creek, making repeated dives, clawing along the mud for the boy's body. After several midstream dives, Fisher surfaced and shouted to the watchers on the bank: 'I've got it.' He had a grip on Lester's body and struck out for the nearer shore opposite the pier, follow-ed by two men in a motorboat. He stood up in waist-deep water near the bank, then staggered, cried out, and dropped into a crouch. From the motorboat, Detective Arthur Van Buskirk saw Fisher with both hands clamped around his right leg – the outside of his thigh, from hip to knee, was missing. Van Buskirk

pulled him into the boat. On the dock men improvised a stretcher from planks and carried him, still conscious, 250 m (273 yds) to the Matawan railroad. He was placed aboard the 5.06 train from Long Branch. At 7.45, as he was being wheeled into the operating theatre of the Monmouth Memorial Hospital, he died. The townspeople – frightened and angered by the monster which had killed a boy just on the verge of leaving childhood, and one of Matawan's most personable young men, in one unbelievable afternoon of horror – collected dynamite and set underwater charges by the pier. They had two hopes: the blasts might kill the shark and they might force Stilwell's body to the surface.

Just as the charges were ready for blasting, a motorboat roared upstream to the pier with another shark victim. Joseph Dunn, a

14-year-old from upper Manhattan, had been swimming with several other boys off a dock 800 m (867 yds) downstream from the Wyckoff pier when someone ran up with a warning: 'There's been two shark attacks upstream – get out of the water!' The boys struck out quickly for the dock. Joseph, the last out, was on the ladder when the shark seized his right leg. 'I felt my leg going down the shark's throat – I thought it would swallow me,' he said. At first Dunn would not give his name because he was afraid his mother would worry about him. Seriously injured, with much of the flesh below the knee stripped from his leg, he was rushed to St Peter's Hospital in New Brunswick.

While a surgeon cleaned and stitched Joseph's severed tendons and lacerated leg muscles, Matawan Creek boiled and spurted geysers as if a primal force had been let loose

---

CHARMED LIVES

Stanley Fisher was one of very few people to have been injured while going to the aid of a shark attack victim.

It is a remarkable fact that on many occasions people have ventured, unarmed into blood-filled water to rescue victims — often wrestling them from the jaws of attacking sharks — without being bitten themselves. The comments made by John Barrett who rescued a fatally injured man from the surf near Sydney in 1935 are typical: 'I didn't stop to think, but dashed straight into the surf. The water was stained deeply red, but my only aim was to reach that poor fellow ... All the time I was acutely aware that the shark must still be with us and I'm not going to hide the fact that I had "the wind up", but I was determined to get him ashore.'

In 1922 two other young Australians, Frank Beaurepaire and Jack Chalmers also went to the aid of an attack victim near Sydney. The victim died, but the two rescuers received an award and £500 each. Beaurepaire used his prize money to start a business which eventually became a multi-million dollar industry, still thriving today.

between its tranquil banks. As indeed it had. In the grip of anger fuelled by fear, the men of Matawan purchased all the town's dynamite in a few hours. Before the sun set the town was also out of ammunition: hundreds of men lined the creek banks armed with handguns, rifles and shotguns. Those without guns brought pitchforks, knives, boathooks, antique harpoons ripped from living-room walls – even hammers. A small army of newspaper reporters and photographers descended on the creek, while newsreel cameramen filmed the vengeful fury and sometimes added to it – especially large charges pushed white geysers high above the creek for the benefit of the newsreels. Shark sightings, and sightings of shadows, were as common as the clams in the low-tide ooze, especially by lanternlight. With the incoming tide, sightings abounded; with the outgoing tide, escaping sharks abounded. A chicken-wire net was strung across the creek just above Wyckoff Dock, and a strong fishnet across the bridge where Captain Cottrell sighted the black shadow. But two days after the tragedy the orgy of vengeance had yielded nothing, and Lester Stilwell's body had still not been recovered.

While the men who sought relief in action shot and dynamited the creek, others who sought the relief of understanding tried to answer the questions: what sort of shark was it, and why these unprecedented attacks at this time?

Dr A. T. Nichols, at the American Museum of Natural History, believed a single shark was responsible for all the attacks. He thought it was a great white or tiger shark which had got out of the Gulf Stream, could no longer find the green turtles which were its staple diet, and had developed a taste for human flesh after the attack on Charles Vansant.

Some local fishermen blamed the Germans: World War I had reduced the number of passenger ships entering New York Harbour which tossed refuse overboard, so the sharks were naturally looking for an alternative source of food.

Although the *New York Times* loftily opined from the safety of its office on West 43rd Street in Manhattan that 'sharks have a much better right to kill us than we have

One of Matawan's younger citizens does her best to look threatening for the New York newspaper photographers. It is to be hoped that she did not discharge the shotgun while holding in this position. For days after the attacks hundreds of people lined the creek banks.

A charge of dynamite sends a huge geyser of water 7.5-m (25-ft) into the air over Matawan Creek. On the opposite bank newspaper photographers and a movie cameraman record the satisfying blast for a public agog for the latest news on the New Jersey horror.

to kill them', dozens of sharks were being hooked, shot and dynamited. On 14 April, funeral services for Lester Stilwell, whose body had not been recovered, and Stanley Fisher were conducted in Matawan. The same day, a Manhattan taxidermist, who had caught a 2.4-m (5-ft) shark off New Jersey, exhibited two bones found in its stomach – one of them identified by physicians as a boy's shinbone.

Joseph Dunn survived. Fifty-nine days after the attack, he left St Peter's Hospital. His leg would always bear the purple scars, but he was able to walk away.

Lester Stilwell's body was eventually found 100 m (110 yds) upstream from the Wyckoff Dock three months after the attack; it was marked by seven bites. On the same day, President Woodrow Wilson's cabinet met in Washington and devoted much of the agenda to the shark menace. The Treasury Secretary promised to instruct the Coast Guard to 'use every means for driving the sharks away or killing them', although he acknowledged that the Coast Guard 'really couldn't do too much about man-eating sharks anyway'.

The Director of the Museum of Natural History, Dr Lucas, offered newspaper reporters the insight that the reason for the attacks was that '1916 is a shark year, just as we have butterfly years and army-worm years.'

Three days after Lester Stilwell's body was found, the mayors of 10 New Jersey coastal resorts issued a statement protesting about news reports which 'cause the public to believe the New Jersey seacoast is infested with sharks, whereas there are no more than any other summer'.

The resort business, the mayors said, had been 'hurt without cause' by the news focus on sharks. Their towns lost an estimated $1 million in holiday cancellations.

QUEST FOR A CULPRIT

There was a lot of conjecture, both at the time and later, about which species of shark was responsible for the attacks at Matawan.

Experts at the Museum of Natural History in New York thought that either a great white or tiger shark was to blame. Indeed, a 2.6-m (8.5-ft) great white shark was caught in Raritan Bay, two days after the attacks at Matawan, with human flesh and bones in its stomach. Clearly it had eaten parts of at least one person, but whether it was Charles Vansant, Charles Bruder, Lester Stilwell or Stanley Fisher could not be determined.

When the wounds on Stanley Fisher's leg were examined and measured it was found that the distance between teeth on opposite sides of the shark's jaws was about 35.6 cm (14 in). Comparisons seemed to indicate that a shark with jaws that size was probably longer than the 2.7-m (9-ft) specimen that witnesses reported seeing in the creek.

The puzzle is that great white and tiger sharks rarely venture into fresh water. The one shark that is commonly found in fresh water, however, is the bull shark *Carcharhinus leucas*. Bull sharks have been found in the Mekong, Zambesi and Mississippi Rivers, as well as in Lake Nicaragua and as far as 4600 km (2600 miles) from the mouth of the Amazon.

NOT ALL SHARKS ARE KILLERS

Few people have a kind word to say about sharks – to most they are simply voracious and indiscriminate killers. Yet not all experienced divers share this view. Here Valerie Taylor, who has more experience with sharks than most divers, attempts to balance the popular image.

With graceful ease a large tiger shark *Galeocerdo cuvier* snaps a bait fish in half and swallows it in one gulp. This 3-m (10-ft) specimen carries the distinctive markings that gave the species its common name. Tiger sharks are most prominently marked when they are very young, after which the patterns gradually fade, becoming almost invisible when they are older. Tigers are oceanic sharks and are found worldwide in warm seas.

Cameraman Ron Taylor (far right) films a tiger shark during the making of the New Zealand feature film *The Silent One*. In the foreground his assistant shepherds the shark back towards the camera. Sessions like this had to be kept brief. The shark soon became tired and distressed and would need to rest before work could continue.

'Except for the tiger shark, I was alone in the lagoon. It was Sunday, my day off, and I was taking stills for my photographic library. I floated against the steel mesh fence that separated the lagoon from the open water. The shark had passed me twice in 30 minutes. Each time I had managed to take two photographs. It was going to be a long day, but the water was warm and some good tiger shark shots would be well worth the effort.

'I first met this particular tiger shark late in the afternoon of 10 October 1983. He was fat, 3.5 m [11.5 ft] long, and no doubt bewildered by what had happened to him. He had been caught some days before by a local diver, and the hook that had penetrated his cheek had been removed. The night of his capture he ate two large tuna, proving that he was in good condition and that his confinement in the lagoon had not spoilt his healthy appetite.

'In the past, Ron (my husband) and I had rarely worked with tiger sharks, mainly because they are not easy to find. Our modest experience suggested that they were calm, easy-going, and not threatening to divers, even when there was food in the water. They were also unafraid of humans, which to anyone not used to sharks can be very unnerving indeed.

'This shark was to appear in a major New Zealand feature film called *The Silent One* which told the story of a young deaf boy who had a giant turtle for a friend. We were working in a lagoon between two coral islands, and by using food, had attracted many marine animals into the area. The shark was needed to add an element of drama and excitement to the plot.

'Using tiger sharks in a feature film is not new: Hollywood is an old hand at it. However, Hollywood sharks always appear to be drugged and half dead, which in fact they usually are. We have never drugged or damaged any shark we wanted to work with, even though critics in the past have wrongly accused us of doing so. Sharks are intelligent creatures, and if coaxed with food quickly develop an understanding of what is expected of them – an understanding which lasts just as long as their hunger.

'Ron insisted that our shark should not be tampered with and the film's producer agreed. The actors and divers quickly became accustomed to working with the shark around, and eventually became quite fond of him. The shark endeavoured to ignore all human activities. Although the lagoon was large and full of natural marine life, he only thought of one thing – escape. A creature of the open water, he circled the lagoon constantly. That is how I knew I would see him every 15 minutes or so, simply by staying against the steel mesh. The beast knew freedom lay just beyond the slender wire and always passed it slowly, searching constantly for a way out.

'I had been in the water some 80 minutes, when I noticed a large sole trying to squeeze through the spaces in the wire mesh. It tried everything, curling up its sides,

The warm tropical and subtropical waters of the Pacific Ocean – with habitats ranging from coral reefs to deep water – are home to many dozens of species of sharks. Valerie Taylor's encounter with a tiger shark, described here, occurred while filming in a lagoon on the fringing reef near the island of Aitutaki, in the Cook Islands, about 2700 km (1680 miles) north-east of the North Island of New Zealand.

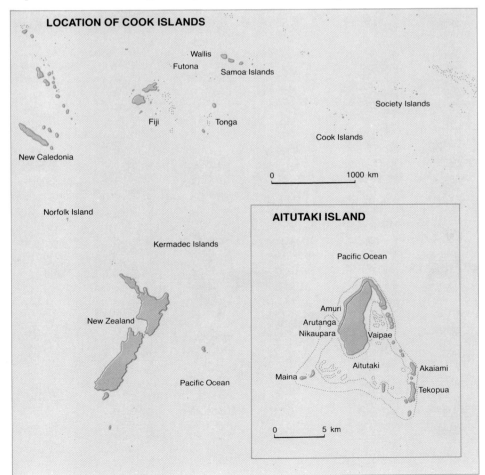

LOCATION OF COOK ISLANDS

Wallis
Futona
Samoa Islands
Society Islands
Fiji   Tonga
Cook Islands
New Caledonia

0   1000 km

Norfolk Island

Kermadec Islands

New Zealand

Pacific Ocean

AITUTAKI ISLAND

Pacific Ocean

Amuri
Arutanga
Nikaupara   Vaipae

Aitutaki   Akaiami
Maina
Tekopua

Pacific Ocean

0   5 km

corner to corner, and finally every square all the way to the surface. Eventually it sunk to the sand and lay with its nose against the mesh. Feeling sorry for the fish, I swam down and began to scoop away sand from the base of the wire.

'Immediately it took a great interest in my efforts. Eventually I had an opening 80 mm [3 in] deep and about 500 mm [20 in] long. The sole immediately swam under and away. I returned to my post just as the tiger reappeared. He passed, black eyes missing nothing. I took my usual two shots. Suddenly, when he was about 5 m [16 ft] past me, the shark stopped, swung back fast and zoomed down to my excavation under the steel mesh. To my amazement, the great fish gently pushed his blunt snout into the gap, and bracing against the bottom, tried to lift the barrier. Three times he tried, each time exerting greater effort without the desired result, then, as though nothing unusual had taken place, the fish continued

patrolling the boundaries of his temporary prison.

'While working with our tiger shark we treated him gently at all times, but still he became distressed if forced to perform for too long before the camera, and would have to be given a rest. Eventually, when the filming had been completed, he was returned to the ocean unharmed by his experience as a film star. For Ron and I his behaviour confirmed what we have known for many years: that sharks, even the so-called dangerous ones, are far more intelligent than is generally believed. If handled in the right way, they will perform in a manner that shows that they have at least as much ability to comprehend a situation as many land animals do.

'My first experience with a tiger shark had taken place some years before, in a channel near Heron Island at the southern end of the Great Barrier Reef. I was snorkelling, gathering shells with

Ever watchful for a way out, the captive tiger shark cruises slowly around the perimeter of his temporary cage. The wound made by the hook used to catch him can still be seen at the corner of his mouth. The tail of a recent fish meal protrudes from one of the shark's gill slits.

two other divers in 5 m [16 ft] of water. It was a beautiful day, and we had been out for about an hour when I sensed something swimming very close to me. Thinking it was one of the other divers I looked around – straight into a large black eye. Little electric shocks seemed to shoot through my body, but even as I swung away I knew instinctively that the shark was not going to harm me. If it was going to bite, it would have done so before I saw it. The fish flowed gently past, huge, beige-coloured, silent. I looked around for the other divers and all three of us swam up over the coral into about 1 m [39 in] of water, and headed towards the island. There was no panic, although my heart was beating wildly. When

A small whitetip reef shark *Triaenodon obesus* searches for food in a tidal pool on the reef at Heron Island – part of Australia's Great Barrier Reef.

confronted by such a creature all defensive action seems useless.

'Back on the island I discovered that the shark was a female tiger, about 4.5 m [14.8 ft] long. She had often been seen by fishermen, and had even helped herself to a few hooked fish. She was probably simply passing by, sensed something unusual in her territory and had cruised over to investigate. This particular shark may have been caught two years later by a fisherman who frequently laid set lines in Shark Bay on the southern end of the island. Now all marine animals, including sharks, are protected around Heron Island, so perhaps one day another big tiger shark will move in to thrill visitors to the reef.

'When I first began diving, 30 years ago, fear of shark attack was my constant companion. The media with its lurid and often inaccurate accounts of shark attacks had conditioned me, along with everyone else, into believing that sharks were monstrous killers. 'The only good shark is a dead one', was the popular saying, and whenever I encountered a shark I always felt it was either it or me.

'Now I look back in wonder at

TIGER SHARK RECORDS

Record sizes for sharks are difficult to confirm. Estimates of lengths are almost always exaggerated, and the actual method of measuring of captured specimens may vary.

What is believed to be the largest tiger shark taken by a diver – a 4.0-m (13-ft) specimen – was killed by Wally Gibbons near Heron Island, Queensland, in 1963.

The record for a tiger shark taken on a rod and line stands at 4.23 m (13.9 ft), although unconfirmed reports have claimed rod and line records of 5.64 m (18.5 ft) near Port Jackson in NSW, and 6.23 m (20.75 ft) in the Gulf of Panama in 1922.

Two 6.4-m (21-ft) specimens have been captured in shark nets – one near Newcastle, NSW, in 1964, the other off Mackay in Queensland in 1980.

Diver Wally Gibbon with a 3.4-m (11.2-ft) tiger shark he killed using a powerhead.

Tricked into attacking, a whitetip reef shark bites Valerie Taylor's arm during tests of the effectiveness of a steel mesh suit. Only by stuffing the suit arm with pieces of fish, and then waving a bait in front of it, could the shark be tempted to bite.

my stupidity, but unlike today's divers, who can learn in 10 minutes from a book what probably took us 10 years of diving to discover, we had no one with experience to advise us. The Austrian diver Hans Hass was filming sharks in the Mediterranean – he did not even have flippers in those days – but we thought that he was crazy and would probably be eaten before too much longer. In Australia Ron Taylor was filming sharks in black and white for Movietone News theatrettes. The most popular footage always showed sharks being hunted, and the audience would cheer every time a shark was seen threshing its life away at the end of a barbless spear. Divers who hunted sharks were popular heroes, and when my turn came I eagerly joined their ranks.

'My attitude towards the sea and its inhabitants has now changed completely, particularly where sharks are concerned. What I once feared, I now respect. It always grieves me to see sharks caught and killed for sport or fun.

'My experience with whitetip reef sharks *Triaenodon obesus* confirms my belief that most sharks are not generally aggressive towards humans. If they attack there is usually a very good reason for them to do so.

'Whitetips are found throughout the Indo-Pacific region. They have little fear of man, and although not normally considered dangerous, are certainly large enough at 1.5 m [4.9 ft], and have sharp enough teeth, to inflict a nasty bite if they wished to. In the presence of food they can become uncontrollable.

'I know of only two attacks on humans by whitetips. On both occasions the victims were spearing fish. Neither attack was fatal, although one, a bite on the shoulder, required extensive surgery. I was told that the diver was holding a bleeding coral trout to his chest at

the time, so the reason for that attack seems quite obvious.

'My own experience with a whitetip attack occurred while Ron and I were carrying out tests on a chain mail suit [see p. 146] to see if it was effective against a shark attack. To coax the shark to bite I had to stuff fresh tuna pieces under the mesh, and then had to attract its attention by waving a whole fish in front of it. After one good bite, usually with a lot of shaking, each shark we experimented with was generally reluctant to try again. Instead it would nuzzle me, trying to find a gap in the mail.

'Far from being aggressive, the

whitetips, once fed, would hang around like friendly puppies, begging for more food. After a while they became a positive nuisance, eating all the bait we put out and swimming around in front of the camera lens.

'However, it was always easy to chase persistent whitetips away. A few smacks on the nose solved the problem. They are quick to recognise aggressive behaviour, and once they had learnt to associate humans with physical pain they would never come within arm's length again – a fact that potential shark attack victims might do well to remember.'

Hungry sharks can be induced to perform for the camera – like this whitetip about to grab a quick bite from a hand-held bait. Divers who hand feed small sharks can generally rely on their being able to distinguish between them and the food. Accidents usually occur when the shark confuses a potential meal – such as a struggling or bleeding fish – with the diver's body. However, only experience can tell a diver which sharks can safely be allowed within attacking distance.

These young divers are sharing the waters of a tropical lagoon with a large tiger shark – one of the three species that has been blamed for most attacks on people. A well-fed shark seems to pose little danger to humans, and would apparently prefer to avoid having anything to do with them.

THE SHARK AND THE LORD MAYOR

The strange story of Sir Brook Watson's encounter with a shark in Havana Harbour in 1749 owes its fame to a unique painting that he commissioned many years later from the American artist John Singleton Copley. Copley's painting is made even more remarkable by the fact that he had probably never even seen a live shark.

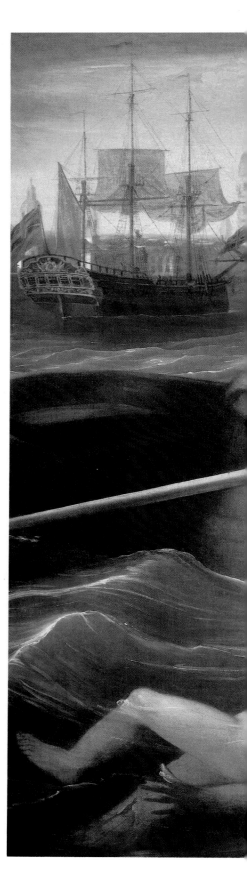

Brook Watson, the orphaned 14-year-old son of an English master mariner, was swimming in Havana Harbour, Cuba, in 1749 when a large shark bit off his right leg below the knee.

Watson's prospects, to say the least, were marginal. Not only were the Cuban surgeons obliged to work without anaesthetic – three burly 'assistants' held amputees down while the saw did its work in those days – but infections were regarded as a matter of course and survival was, by modern standards, little short of miraculous. Watson, however, survived not only the shark attack, but also the surgery. With his new wooden leg, he returned to Boston, Massachusetts, where a distant relative had been taking care of him – only to discover

that the relative was bankrupt. The young Brook moved to Nova Scotia, Canada, and served as a commissary (concerned with supplying provisions) on the British side in the war between British Canadians and the French. He then helped evacuate British loyalists from the United States during the American War of Independence. Brook Watson's head was not filled with dreams of justice and republicanism. He believed in the established order, not in the ideas of visionaries and agitators.

He established himself as a merchant in London and became an able defender of the merchants' faith that progress for humanity lay in trade and commerce, and the laws and order which made commerce both possible and

DETAILS OF A DEADLY BITE

The shark in John Copley's painting most closely resembles a tiger shark, a species found in Havana Harbour. It seems unlikely that Copley ever saw a live shark, or even a complete dead specimen, but he may have had a set of shark's jaws to use as a model. The director of the New England Aquarium, John Prescott, has suggested that the finished work is a combination of features from various sharks, with two kinds of teeth and imaginary lips.

The lips on Copley's shark are interesting when compared to the mouth of a live attacking shark. When a shark, such as a tiger or a great white (bottom right), is about to attack it changes the whole shape of its head. The head and snout are lifted, and the upper jaw is pushed forward so that it protrudes from the mouth. Once the prey is securely impaled on the teeth of the lower jaw, the upper jaw closes and the shark swings its head and the front of its body violently from side to side, sawing the sharp upper teeth, which are often serrated, through the flesh until a large chunk has been gouged out.

Copley's curious shark in close-up.

A great white shark seizing a bait.

John Singleton Copley's dramatic reconstruction of the awful events in Havana Harbour in 1749 was in fact painted almost 30 years after it occurred. The work was exhibited at the Royal Academy in London in 1778 where it created a minor sensation and lively debate over its accuracy.

profitable. In 1779 he took a leading part in the formation of the Light Horse Volunteers who helped to suppress the riots fuelled by the fury of the dispossessed poor in London in 1780.

He was elected to the House of Commons for the seat of the City of London in April, 1784 and held the seat for nine years, becoming in the meantime a director of the Bank of England. He resigned the seat in 1793 to become commissary-general to the Duke of York's army in Flanders, and he served with the army until his return to England in 1795.

The wooden leg became Watson's symbol. In the Britain of the 1790s, the image of the bold adventurer, savaged in childhood on the exotic Caribbean island of Cuba by the tiger of the deep, was politically potent, and Watson exploited it to the full.

He was elected Lord Mayor of London in November 1796 – the culmination of a career with a fairy-tale quality to rival that of Dick Whittington. When visitors to the Lord Mayor's office had not heard the story of the most famous wooden leg in Britain, he delighted in mystifying them, saying baldly in response to questions that the leg 'was bit off'.

As part of his exploitation of the shark attack, Watson commissioned the American artist John Singleton Copley in 1778 to paint the incident. Copley, a self-taught Boston artist, moved to London in 1775 to escape the artistic provincialism of New England, and the trouble being stirred up by those notorious agitators George Washington and Thomas Jefferson. The American War of Independence broke out a year later, in 1776.

Copley's painting, which one critic said 'stands alone in its age', tells a vividly symbolic story. Its broad theme, the struggle between mankind and the wild, is as old as art itself. But Copley's painting had

Watson was made a baronet four months after this engraving of him was published in a London journal. He was 68 years old at the time. He lived only for a further four years and died at East Sheen, Surrey, on 2 October 1807. A contemporary obituary described him as: 'a diligent, zealous and faithful servant; a firm, upright and merciful magistrate; to his wife, a most affectionate and tender husband.'

Watson's remarkable coat of arms was granted in 1803 at the time he was made a baronet. He specially requested 'such Arms as may contain an allusion to an awful event in his life, and be a memorial of his gratitude to Heaven for his signal preservation on that occasion.' The letters patent granting the Arms duly specified, among others, these features: 'a human leg erect and erased below the knee' and a gold-crowned Neptune 'repelling a shark in the act of seizing its prey proper'. The Latin motto, *Scuto Divino*, could be loosely translated as meaning 'Under God's protection'.

broader themes: art critics saw in the figure of Watson a neoclassical allusion, man as eternal being, and for some, the sailors reaching towards Watson were like the images of Christ's disciples with a fishing net in a famous Raphael tapestry entitled *The Miraculous Draught of Fishes*.

The contrast between the anguish and terror on the boat, the furious energy of the thrusting spear and the deadly movement of the shark with open jaws, with Watson's pale body floating languidly in the face of primal terror focuses attention on the 'man versus beast' theme – and on Watson as larger-than-life. When the painting was exhibited it caused an uproar. Nit-picking critics questioned the accuracy of some nautical details, and wondered about the authenticity of the shark. However, despite the fact that Copley had never visited Havana, his background details of the harbour and town are historically accurate. But the *St James Chronicle* was not among the nit-pickers. Copley's painting, the *Chronicle* said in 1778, showed he was 'a Genius who bids fair to rival the Great Masters of the Ancient Italian Schools.' He was elected to full membership of London's

prestigious Royal Academy the following year.

Outside the art world, not all were lost in wonder at the portrayal of a wealthy, powerful Tory politician as a mythical figure of heroic proportions. After the 1784 election, a satirical series in the form of reviews of an imaginary epic called *The Rolliad* were published by reformist Whig writers. One of them, John Wilkes, was unkind enough to suggest that had the shark torn off Watson's head instead of his leg, a wooden head would have done as well for him as the wooden leg had.

The verse satirised both Watson's famous accident and his manner of speech:

'One moment's time may I
   presume to beg?'
Cries modest Watson, on his
   wooden leg;
That leg; in which such
   wond'rous art is shown,
It almost seems to serve him
   like his own;
Oh! had the monster, who for
   breakfast eat
That luckless limb, his noblest
   noddle met,
The best of workmen, nor the
   best of wood,
Had scarce supply'd him with a
   head so good.

POWERFUL SYMBOLS

Almost every conceivable object has been used on a coat of arms at one time or another, and both dogfishes and sharks have occasional appearances.

The families of Jessie and Harrie both have three dogfishes on their arms (left), and several English and Irish families have adopted the device of a shark's head swallowing a man (right). A curious fastidiousness made the creator of the arms show the unfortunate victim as a negro – apparently to avoid offending European sensibilities.

It is perhaps hardly surprising that sharks have not been popular on national flags. One notable exception, however, is the Solomon Islands which became independent on 7 June 1978. On that occasion a new green and blue flag was flown for the first time which featured two frigate birds, a crocodile and a shark.

Keeping Sharks At Bay

Is there any really effective protection from shark attack? Jim Stewart, at the Scripps Institute of Oceanography, and himself a shark attack victim, has a terse answer to this question: 'Yes, the shade of an oak tree'. The only way to avoid shark attack is to stay away from the water.

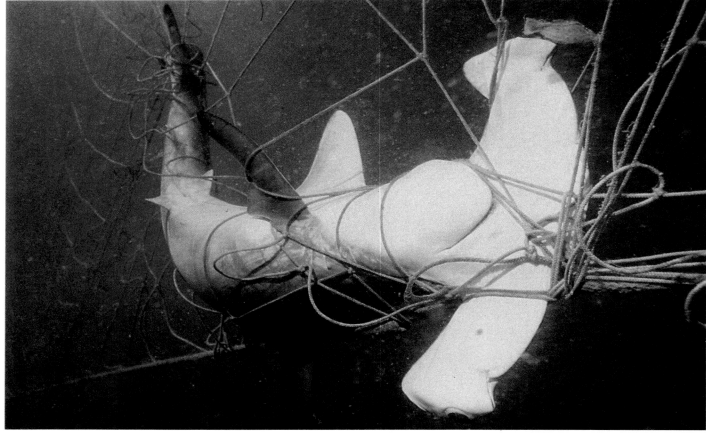

Swimmers and bathers are the most frequent victims of shark attack and a great deal of effort has been expended over the years in protecting beaches. One way to do this is simply to build a fence around the area. This has been done in South Africa, Australia, the Panama Canal zone and in other places around the world. While unsightly, these structures are effective if properly maintained. The reason they are not used more is that the initial costs are high, and so is the cost of maintenance and repair. The pounding that the fences receive from the ocean quickly damages them, and they must be constantly inspected for holes and gaps.

When Richard Nixon was President of the United States he had a house on the Atlantic Ocean near Miami, Florida. His private bathing beach was protected by a fence. One day the Secret Service was embarrassed to find sharks swimming inside the enclosure. It turned out that there was no plot against the president's life, it was simply that a hole had developed in the fence. After the hole was fixed and the sharks removed, navy frogmen were given the task of inspecting the barrier.

A system of protection used in

A large scalloped hammerhead, entangled in a meshing net off a Sydney beach. The numbers of sharks caught in these nets has declined in recent years.

Australia and South Africa simply involves catching as many sharks as possible in popular bathing areas, using a method called meshing. Meshing consists of setting nets at intervals along beaches. The nets are set in deep water parallel to the beach line. Sharks swimming either towards the beach, or away from it, usually at night, become entangled in the nets and die. By drastically lowering the number of sharks the chance of attack is greatly reduced, and this method has proved to be

Shark-proof enclosures, provided they are properly maintained, offer the best protection against attack. This enclosure is at a beach in Sydney Harbour where it is not exposed to the full force of ocean waves. Enclosures on ocean beaches are generally quickly damaged or destroyed by large seas.

Shark fishing contests, which offer large prizes to successful anglers, have been effective in controlling shark numbers in some parts of the United States of America. However, some researchers fear that the large-scale slaughter of sharks will permanently affect populations of rarer species.

quite effective. The disadvantages of meshing are the relatively high cost involved, and the large number of harmless sharks, large fish, and even marine mammals that are killed unintentionally.

Along the coast of the United States bordering the Atlantic Ocean and Gulf of Mexico, shark fishing tournaments are becoming very popular, and this has proved to be an inexpensive method of reducing shark populations. Entrants pay a fee of between $25 and $100 for the chance to win total prizes of up to $50 000. The illegal, but inevitable, side bets can run into hundreds of thousands of dollars for the person who catches the largest shark.

Sharks have always been exploited for food in the Orient, and now their popularity is slowly spreading to other parts of the world. The number of sharks landed by fishermen in the United States has increased dramatically since 1980. Surprisingly this interest in sharks as food fish has not spread to Australia and New Zealand. The meat of the great white shark is exported from New Zealand to the United States where it is considered a delicacy rivalling swordfish.

Shark fins were being sold for about $(US)16 per kilo (2.2 lb) in 1985, and a bowl of shark fin soup in a restaurant could command up to $(US)20. Shark meat is lower in cholesterol than any other fish, and restaurant servings of great white and mako flesh brought $(US)3.50 per 100 g (3.5 oz). Apart from being used for food, sharks also provide other products which are much in demand. A set of large shark jaws may fetch $(US)400. Leather made from shark skin is more durable than cow or pig hide, and boots, wallets and other items made from shark leather command high prices. Shark-eye corneas have been used for human transplants, and shark tissue to make artificial skin for burn victims.

Research on electrical shark barriers has been conducted in South Africa for more than 20 years. In this system current-carrying cable are located off beaches, outside the surf zone. Pulses of electric current are passed through the cables creating an electric field in the water. If the field is strong enough, sharks will not swim across the cables, and are repelled. This system has been successfully tested, but has not been put into use because of the high cost of operating and maintaining the equipment. Research is continuing, however.

Another system for repelling sharks that was widely publicised was the bubble-curtain. To produce a curtain or wall of bubbles a perforated length of hose was laid at the bottom of a tank and air was pumped through it. However, Dr Perry Gilbert and his colleagues tested the bubble curtain on 12 large tiger sharks at the Lerner Marine Laboratory in the Bahamas. They found that 11 of the sharks swam freely through the curtain of bubbles; only one of the sharks was repelled by it. These results point up one very important feature of shark countermeasure tests. Tests must be done, not only on many shark species, but also on a large number of the same species to ensure that the results are accurate and consistent.

Experiments using sound to repel sharks have been conducted both in Australia and the United States. Very intense sounds were found to frighten sharks away

---

A SUCCESS STORY

Meshing was first introduced in Sydney in 1937. In the 18 years before that date there had been 14 attacks on Sydney's 20-odd ocean beaches, and 15 inside the harbour. Since 1937 there has not been a fatal attack on any ocean beach, although there have been some incidents. The number of attacks in the harbour has been reduced to seven.

The meshing is carried out by contractors who lay nylon gillnets 150 m (490 ft) long and 6 m (20 ft) deep, parallel to the beach and about 300 to 500 m (330 to 550 yards) offshore. Each net is suspended from floats and is anchored to the bottom by lead weights. Nets must be laid at each beach at least four times a month, and they must remain in place for a minimum of 24 hours (48 hours at weekends). It should be stressed that the net does not cover the entire length of the beach, nor does it stretch from the surface to the sea floor.

The success of the operation can be gauged by the number of sharks removed from the nets. Seven hundred and fifty-one were caught in 1940, and by 1948 this had declined to 260. In 1975/76 only 362 sharks were caught off Sydney, Newcastle and Wollongong (two nearby coastal cities) combined. In 1980/81 this figure had dropped to 163.

A diver embraces a large grey nurse he has just killed with the powerhead in his right hand. This type contains a shotgun cartridge which discharges when it is pressed against the shark's body. Other types use a .303 rifle bullet, and are fired at the shark from a rubber-powered spear gun.

under some circumstances, but the results have not been consistent.

Many methods have been developed to protect divers from shark attack; some work very well, others are only marginally successful at best. First among these are the various anti-shark weapons. Spear guns alone are not very effective, but if the spear is tipped with a powerhead then this combination (sometimes called a bang stick) can be quite deadly. A powerhead is simply a single-barrelled gun of almost any calibre (.303 is popular in Australia) designed to fire when it comes in contact with the shark. One par-ticularly successful version uses a small explosive charge instead of a cartridge. While effective, these weapons require a good deal of skill on the part of the diver. A shot to the brain is usually necessary for a quick kill.

Gas injection darts were very popular a few years ago, and are still available in some diving supply shops. These weapons work by injecting carbon dioxide gas into the shark's body cavity through a large, hollow needle, causing various problems for the shark. Sharks are about five per cent heavier than the water they are swimming in, so a gas bubble inside their bodies makes them bouyant and forces them to the surface. Larger amounts of gas stiffen their bodies to the point where they cannot swim at all. The main difficulty with these devices is that the gas needle must penetrate the shark's body cavity, and this can only be done from the side or underneath the animal. Head-on shots are not effective. Of course, such weapons are of little value if the shark is not seen first, and in about half of the attacks on divers this was the case.

Several electrical shark repellers for divers have been developed. These are either carried by the diver, or the electrodes are

incorporated into the diver's wet suit. Electrical repellers work, but have never been widely sold be-cause of the high cost of manu-facture and the limited market.

The June 1974 edition of *Skin Diver* magazine contained an article entitled *Revolutionary Weapon: Anti-Shark Wet Suit*. This turned out to be a black wet suit painted with white stripes. The idea for this unusual suit arose from the observations that pilot fish, which often swim with sharks, have vertical black and white stripes, and that some venomous sea snakes have black and white rings around their bodies. United States Navy scientists tested the suit with blue sharks and grey reef sharks, among others, and found it completely ineffective. Ron and Valerie Taylor, the famous Australian underwater

film makers, have also confirmed that the striped suit does not repel grey reef sharks (see p. 82).

Several attempts have been made to design shark-proof suits. The basic problem with such designs is that they must allow the diver ample mobility, while at the same time providing adequate protection. Since most attack victims suffer cuts that cause shock and bleeding, the main objective is to prevent the shark's razor sharp teeth from penetrating. Severe bruises and broken limbs are not life-threatening, and can be tolerated. This is the sort of compromise that must be made in order to keep the weight and flexibility within acceptable limits.

The United States Navy investigated a material named Kevlar which is widely used in bulletproof vests. Unfortunately, while the material may stop bullets, it did not stop sharks. Kevlar performed well against blue sharks, but failed when tested against the more powerful jaws of large lemon sharks *Negaprion brevirostris*. However, the material is now being used in suits worn by salvage divers to protect them from jagged metal edges and other sharp objects.

A very effective suit has been developed by Ron and Valerie Taylor, together with an American diver, Jeremiah Sullivan. It is simply a light-weight stainless steel mesh suit, similar to those worn by medieval warriors. The material employed has interlocking metal rings, and is widely used in the manufacture of butcher's gloves, aprons and arm protectors. Many tests have now been made in which the Taylors and Sullivan have deliberately allowed themselves to be bitten while wearing the suits. No injury, except for minor bruises and scratches, has been sustained (see p. 142).

The only shortcoming of these shark suits is their cost. A full suit may cost $(US)5200. While this is

This Kevlar suit is one of several attempts that have been made to find a material that will protect divers against a shark attack. So far no material, other than steel mesh (see p. 142), has successfully repelled the sharp teeth and powerful jaws of a large shark. Even if a suitable material is found, divers will still have to accept the possibility of serious bruising or crushing injuries that a large shark will inflict when it bites.

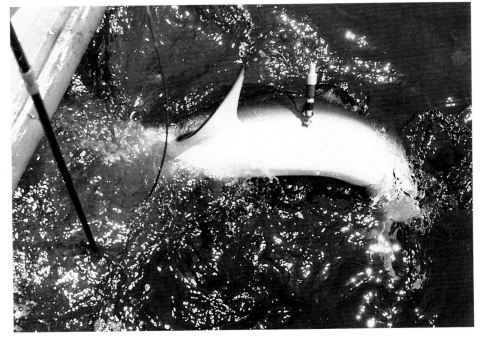

A blue shark, its body distended, struggles against the effects of carbon dioxide gas injected into it from the dart in its side. Divers attempting to use such a device in self-defence will need to be certain of their aim, because the gas is only effective if it enters the body cavity of the animal.

A **United States Navy** researcher (left), wearing a life jacket, scoops water into a Shark Screen to test its effectiveness. Three inflatable collars which are blown up by the user keep the bag afloat, and these can be brightly coloured to make them obvious from the air. Beneath the water (above) all that can be seen is a dark, shapeless mass which has no attraction whatsoever for sharks. The plastic also prevents blood or urine from escaping into the surrounding water.

out of the reach of most sport divers, professionals and others who must swim with sharks would be well advised to buy one. The high cost stems from the method of manufacture. The mesh must still be made in the same way that it was in medieval times – each of the steel rings has to be linked together individually and welded by hand.

A problem that has received a good deal of attention, especially from the armed services, is that of protecting the survivors of air crashes and sinking ships from shark attack. Unfortunately, the problems of finding and rescuing castaways, and protecting them from attack, can conflict. In order to help make a person visible in the sea it is best if they are wearing highly visible clothing. But sharks are more likely to approach brightly coloured objects than darker ones. For many years the flight crews of various armed services wore, and

still wear, bright international orange – dubbed 'yum yum yellow' – flight suits. During the Vietnam War the United States military changed the colour of flight suits to dark green to make them harder to see on both land and sea. However, life rafts and life jackets continued to be orange or yellow. Bright colours that speed rescue are definitely preferred because the chances of shark attack are extremely small. In the past 20 years there has not been a shark attack reported on a United States serviceman on active duty, and there has only been one since World War II. This is in spite of the fact that there have been many hundreds of occasions when servicemen have had to parachute into the oceans of the world. In cases where the individual did not survive, but the body was recovered, none has shown evidence of shark bite. For this reason little research

on shark deterrents has been done in the United States recently.

During World War II, when large numbers of men and women were being forced into the water from sinking ships and other casualties of war, an urgent programme to find a chemical shark repellent was initiated. This hastily conducted research resulted in the development of the Shark Chaser, a solid 127-g (4.5-oz) packet of chemicals – copper acetate and a nigrocine dye – which was attached to the life jackets of servicemen. Tests conducted after the war proved Shark Chaser to be completely ineffective as a shark repellent. Its use was discontinued in 1976. Although useless, Shark Chaser probably provided peace of mind for many unfortunate individuals floating in the sea and fearing shark attack.

About 15 years ago the United States Navy developed and tested a

Orange is the most popular colour for flight suits because it is easy to spot from the air during a search. It is an unfortunate coincidence that this is also one of very few colours that have been shown to actually attract sharks.

Urgent research into shark repellents during World War II lead to the development of the Shark Chaser – a packet containing copper acetate and nigrosine dye. Its major contribution was to give servicemen peace of mind.

device for protecting people floating in the sea. It is called the Shark Screen, and consists of a large plastic bag which is closed at the bottom and has inflatable collars at the top. When not in use it is folded into a small packet 76 x 114 x 178 mm (3 x 4.5 x 7 in), which weighs about 0.45 kg (1 lb). To use the device, it is unfolded and one or more of the three collars is inflated. The user then climbs into the bag and scoops water into it until the device is completely extended. Once inside, the Shark Screen user is completely hidden from view – all that can be seen from under the water is a large, dark, bulbous shape which sharks are reluctant to approach, even when fresh fish are attached to it. All blood and other potential chemical attractants are kept inside. Even electric fields that may encourage an attack are screened, since the device is made of an insulating material.

Sharks that penetrate the Moses sole's brilliant camouflage get a nasty surprise.

NATURAL PROTECTION

Break Through! 100 per cent Effective Shark Repellent was the title of an article in the September 1975 issue of *Oceans* magazine. Unfortunately, subsequent testing of the repellent has lowered the effectiveness considerably below 100 per cent. The article referred to the discovery by Dr Eugenie Clark that a small, flat fish called the Moses sole *Pardachirus marmoratus* secreted a milky fluid that discouraged sharks and other large fish from eating them. The problem with this substance, and with all the other 200 or so chemicals that have been tested so far, is that a considerable quantity is needed to make a practical repellent. Chemicals simply disperse too rapidly in the ocean. To use the Moses sole poison as an effective repellent would require that nearly one kilogram (2 lb) of the substance be released continuously per hour. The reason it works for the Moses sole is that the fish releases the toxin directly into the shark's mouth, thereby producing very high concentrations of the chemical, while expending very little of it at any one time.

Despite the fact that initial results were discouraging, work on the Moses sole toxin was still continuing in 1985, when researchers at the Hebrew University in Jerusalem announced that the active constituent resembled household detergent.

THE SUIT OF STEEL

A suit that would give divers protection against shark attack is an attractive possibility. Several attempts have been made to develop one, but the problem is always to keep the garment light and flexible enough for practical use. Only the steel mesh suit has had any real success so far.

The idea for the shark suit came to diver and photographer Ron Taylor in 1967 while he was working with a Belgian scientific expedition on the Great Barrier Reef.

One of the expedition members had with him a pair of gloves made from steel mesh to protect his hands from cuts and scratches. These had originally been made for meat workers to wear while boning carcases – protection from the razor-sharp knives is essential. The first suit was made to Ron Taylor's

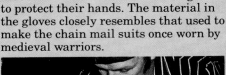

Meat workers wear metal mesh gloves to protect their hands. The material in the gloves closely resembles that used to make the chain mail suits once worn by medieval warriors.

design with the help of Jeremiah Sullivan, a marine biologist studying at Scripps Institution of Oceanography. The finished product weighed some 6 kg (13 lb) – about the same as a diver's weight belt – and was fastened at the wrists and ankles with straps. A hood covered the diver's head, except for the mask.

The dubious honour of being first to try out the new suit fell to Valerie Taylor during a trip to the Coral Sea in December 1978. Nervous at first about the ability of the suit to protect her, Valerie watched as large grey reef sharks circled around, snapping up fish baits as they were released. The sharks clearly knew she was there, but no attempt was made to attack.

The next day, with increasing confidence, Valerie started to handle the sharks – whitetips this time – and even attempted to force her arm into the mouth of one of them. Still they refused to bite, although one did nip her on the leg, but without causing any damage.

Experiments with the suit continued on and off for the next six months until, in June 1980, the prototype received its first real test. On this occasion the Taylors and Sullivan were diving off the

Ron Taylor uses the jaws of a live blue shark to test the suit. While the suit prevents the teeth from penetrating, it offers no protection from the formidable crushing power of a large shark's jaws.

Californian coast with blue sharks *Prionace glauca*, which have no fear in the presence of divers. Experiments showed that the sharks could be tempted to bite by holding a bait in front of the mesh-covered arm, and then withdrawing it at the last moment. Once committed to a charge the sharks seemed unable to change their minds at the last moment. Later Valerie recorded the results of the experiment in her diary: 'I still felt a bit nervous, and like Jeremiah, tended to push the sharks away. Some deep rooted instinct of self-preservation had me defending myself even when I wanted the attack to take place. A big blue shark caught me unawares and latched onto my arm with a sudden thump. I was, to say the least, startled. A natural reaction to seeing a neat set of razor sharp teeth grinding into one's body with mindless fury. The nictitating membrane flicked back for a second as the fish looked at me, an eye black as ebony, and with as much expression gazed into my two blue ones. As the shock wore off, I realized that there was no blood and that it wasn't really hurting. There was just the initial thump and a squeezed feeling. Being pulled back and forth by my elbow was uncomfortable, but while it looked agonising, it wasn't.'

After the initial experiments, the suit was tried on many other occasions, and with several different species of sharks – always with the same success. Plans were made to extend the tests to great whites – the most formidable and powerful of all sharks – but these were postponed. This was not so much because of fears about the power of the shark's bite – which the Taylors believe to be less damaging than is popularly assumed – but because a large great white is quite capable of carrying off both the diver and the suit in one piece.

A two-metre (6.6-ft) blue shark attempts to sink its teeth into Valerie Taylor's arm. Shortly after this photograph was taken another shark bit Valerie's unprotected leg, inflicting a wound that needed hospital treatment.

SHARK FOR SALE

The promise of riches has tempted many enterprising businessmen into investing in shark fisheries, but few have succeeded in making their fortunes from them. Apart from shark flesh, which is widely eaten, only fins and hides are now sold in commercial quantities.

The history of shark fisheries is one of boom and bust. Most species are slow to reproduce, and once stocks are reduced beyond a certain point they take a long time to build up again. In the 1930s and 40s soupfin sharks *Galeorhinus galeus*, then found in huge numbers off the coast of California, were in great demand for the oil their livers contained. In 1939, the fisheries' peak year, over 4 million kg (9 million lb) of sharks were landed. By 1944 the catch had fallen to only 270 000 kg (600 000 lb) as stocks were virtually wiped out. Today, over 30 years later, their numbers have still not recovered.

Sharks are eaten in many parts of the world, although in some countries the flesh has never been popular, except under a pseudonym. Distaste for the idea of eating them probably stems from the fact that sharks consume any unpleasant rubbish thrown from ships. Sailors have always been reluctant to eat them. In 1771 Peter Orbeck, travelling between Europe and the East Indies, described attitudes aboard his ship: 'If a sailor dies in a place where dog-fishes haunt, and is thrown overboard, he is sure to be buried in the bellies of some of them. Large dog-fishes are never eaten, and small ones but seldom, and in cases of necessity only. They are cut into slices, which are squeezed in water till no oil remains in them; after being thus washed, the flesh is boiled or roasted: the part towards the tail is the best; the fore-part is seldom eaten.'

These days many consider the

TREASURES FROM THE SEA

Hides, oil and fins are the three most valuable products (apart from flesh) obtained from sharks. Today only fins (see p 146) and hides are still exploited on a commercial scale. The shark oil industry declined in the 1950s when vitamin A was first synthesised. Until then shark liver oil had been a major source of the vitamin, especially during World War II when cod liver oil supplies were interrupted.

Shark leather has always been in demand because it is particularly tough, although products made from it are usually expensive because the skins are difficult to obtain and to process. The curing and tanning stages are essentially the same for shark skins as for more conventional hides, but shark skins have the added problem of being covered by hard, abrasive denticles. These must be removed from the surface – in a process known as de-armouring – by soaking the skins in acid or lime.

One of the most successful shark leather manufacturers is the Ocean Leather Corporation of Newark, New Jersey in the United States of America. First established in 1921, the company was originally set up to exploit sharks for oil, hides, flesh and fertiliser. Soon, however, it ran into the same problem that confronts most shark-based enterprises – concentrated fishing rapidly reduces the shark population to a point where it is no longer economic to exploit them. Ocean Leather now concentrates on producing shark leather, and handles about 50 000 skins a year. The majority of the skins come from large sharks, such as tigers, caught in waters off the Pacific coast of Mexico and Costa Rica. Most of Ocean Leather's output is turned into shoes, handbags, wallets and belts.

Shark hides being inspected at Ocean Leather's Newark warehouse.

Shark products on display in a Sydney museum in the 1940s, at a time when sharks seemed to offer considerable scope for exploitation. Now only edible sharks are fished commercially in Australian waters.

Equipment for extracting oil from shark livers at an Australian plant in the 1930s.

Jewellery made from shark teeth can fetch high prices. The white tooth is that of a great white shark, and the brown one a fossil tooth from the extinct shark *Carcharodon megalodon*.

tastiest shark to be the mako *Isurus oxyrhinchus*, and the flesh has been compared to tuna by enthusiasts. In some countries it appears on menus as 'calf fish' or 'sea sturgeon'. The related porbeagle shark *Lamna nasus* and great white shark *Carcharodon carcharias* are also highly regarded.

The most widely eaten species – and probably the most abundant shark in the world – is the piked dogfish *Squalus acanthias*, which masquerades as 'rock salmon' in Britain and 'flake' in Australia. It is heavily fished in American, Canadian, New Zealand, Japanese and Korean waters, and an estimated 34 000 tonnes were caught in 1978.

Another shark that is caught in large numbers for food is tope (also known as the soupfin shark or school shark) *Galeorhinus galeus*. This was the shark that supported huge fisheries off the Californian

AN ANCIENT RECIPE FOR SUCCESS

Shark fins are used in a number of Chinese recipes, although it is shark fin soup which is best known to European gourmets. Various versions of this oriental delicacy are available in restaurants around the world.

All of the four main fins on a shark – pectoral, dorsal, anal and tail – are used for soup making, although it is the lower tail fin that is particularly highly prized. The fins are cut from a shark, trimmed to remove any flesh, and left in the sun to dry for about two weeks. Once dry, the parchment-like fins are ready for shipping, and it is in this form that they are usually sent to China for further processing.

The next stage in the preparation of the fins involves boiling them for several days until the fibres within – which look like uncooked noodles – can be removed. It is these gelatinous fibres that are dried and used in soup making.

There are many recipes for shark fin soup, and all involve soaking and cooking the dried fibres for several hours before adding chicken stock and a variety of other ingredients. Most commentators, however, seem to agree that the shark's fin itself has little or no flavour. Perhaps very few have tasted the 'real' soup, which is reputed to take several days to prepare at the hands of a master cook.

A tin of ready-made shark fin soup.

Shark fins drying aboard a fishing boat (left). Vast numbers of these are shipped to China every year to be turned into culinary delicacies.

Dried shark fin (below) in the form in which it is available at most Asian stores. The gelatinous fibres must be soaked in water for two hours, and cooked for another three hours before they can be eaten.

coast in the 1940s, and also a South African fishery at about the same time. Both industries declined because of over-exploitation. Tope (known locally as Australian school shark) has also been fished off the coast of Australia since the 1920s, although the industry has had a number of problems. When stocks declined in the 1940s a minimum size of 91 cm (36 in) was placed on any sharks caught. Then, in the 1970s, high mercury levels were found in some large specimens, so a maximum size of 104 cm (41 in) was laid down. Despite these restrictions the fishery continues, but catches are declining.

Although most sharks are edible, there are a few that are actually poisonous to eat. These include some tropical species, and the Greenland shark *Somniosus microcephalus*.

The toxin in the flesh of the Greenland shark has not yet been identified, but this has not prevented the Icelandic people from eating it, as they have done for hundreds of years. A shark is prepared for human consumption by burying it in a pit near the water's edge for several months while bacteria modify the flesh. After that it is cut into chunks and hung in a barn for about six months to cure. Thin slices of *Hàkall*, as it is called, are then ready to be eaten with a glass of aquavit to wash them down. The taste of ammonia is apparently so strong that some intrepid experimenters have compared it to eating smelling salts.

Shark fins also have the reputation of being an acquired taste. The subtle flavour they impart to dishes – particularly to the famous soup – often eludes uneducated palates. There is a constant demand for fins in China (see box p 146) which provide a useful source of additional income for the owners of commercial long-line fishing boats.

At one time oil was among the most important by-products of shark fisheries, but that industry has now almost completely disappeared. At various times shark oil has been used as a lubricant, in oil lamps, for curing leather, in soap making, as a tonic, as an ointment, as a base for paints and as an important source of Vitamin A. Oil extracted from the livers of spiny dogfish has been found to contain

PROBLEMS OF A PIONEERING INDUSTRY

There have been many attempts over the years to exploit sharks commercially, and almost all have ended in failure. The story of the rise and fall of Marine Industries Ltd, started in 1927 at Pindimar, 200 km (125 miles) north of Sydney, by Norman Caldwell and three companions, is typical.

The sharks were caught in nets which were left in the water overnight, and early hauls were spectacular. On the first day near

Pindimar – in a bay which locals considered to be free of sharks – 30 were caught, ranging in size from 45 to 225 kg (100 to 500 lbs). Shark products brought good prices (see below) and Caldwell reckoned that the hide, fins and oil from a 225-kg (500-lb) whaler shark brought in £3 7s 9d for a total outlay of only £1 (the average wage at the time was about £3 10s a week).

Soon, however, problems began to surface.

The catch in any area dropped rapidly after only one day's netting, and the company's small boat was forced to go furthur and further afield to find sharks in adequate numbers. A bigger boat was needed, but declining catches meant that there was not enough money to invest in badly needed improvements. It was soon impossible to continue operations and in 1932 the Pindimar factory was closed down, never to re-open.

Shark products fetched good prices – it was the decline in shark numbers that eventually forced the closure of Marine Industries, after six year's hard work.

Dorsal fin Dried fins were sold in China where they are treated and used as the basis for a highly regarded soup. Large fins take about two weeks to cure and lose about 40 per cent of their weight. Dried dorsal fins fetched up to 6s 11d per kg (3s per lb).

Skin Tanned hides, with the denticles removed, were turned into shoes, bags and suitcases. Ordinary shark leather fetched £2 16s 6d per sq. m (5s 3d per sq. ft) and carpet shark leather £5 7s 8d per sq. m (10s per sq. ft).

Flesh Dried shark flesh was ground into a meal, rich in protein, which made excellent cattle food. A tonne of dried flesh fetched £20.

Head Dried and and ground shark heads were used as fertiliser, and fetched £12 per tonne.

Liver Cut into lengths, the liver was rendered down in a steam-heated kettle. A 4-m (13-ft) tiger shark liver could yield as much as 82 litres (18 gallons) of oil rich in vitamins A and D. Shark oil had medicinal qualities, and was used for human consumption as well as for stock feed. What was left of the liver was dried and turned into meal, which was also fed to animals. The oil fetched 7d per litre (2s 9d a gallon).

Caudal fins The lower lobe of the tail fin is the most valued for soup making. They lose little weight while drying and fetched up to 11s 4d per kg (5s per lb).

Pectoral fins All fins are used for soup making. Pectoral fins fetched about the same price as dorsal fins.

Anal fin Anal fins gave the poorest yield of gelatinous fibres for soup making. They fetched only 3s 5d per kg (1s 6d per lb) when dried.

ten times the amount of vitamin A present in cod liver oil, which was once the principal source.

Before World War II most of the vitamin A consumed in the United States and elsewhere in the world was imported from Europe. When war broke out and European fisheries were interrupted, there was a great demand for sharks from waters along the Pacific coast of the United States. In the seven years from 1937 to 1943 thousands of tonnes of sharks were caught, and by 1944 they had virtually been fished out. In 1950 chemists discovered a method of making vitamin A artificially, and the demand for shark liver oil disappeared virtually overnight.

However, one chemical present in shark liver oil – squalene – is still the subject of considerable interest. It is an unsaturated hydrocarbon which is found in large quantities in the livers of some sharks and, to a lesser extent, in the livers of most higher animals. Its function is unknown, but at one time it was suggested that it might be effective against malignant tumours, because sharks were thought to be free of them. Unfortunately, studies showed that sharks do have tumours, but the medicinal properties of squalene are still being investigated.

There are extraordinary quantities of oil in the livers of large sharks, particularly basking sharks *Cetorhinus maximus*. In 1945 author Gavin Maxwell set up a basking shark fishery in the Hebrides, the islands off the coast of northern Scotland, which lasted only about three years, but yielded a lot of information about these giant creatures. Maxwell estimated that an 8.8-m (29-ft) shark, weighing around 6.5 tonnes, contained a 940-kg (2072-lb) liver which might yield 2270 litres (500 gallons) of oil!

There has been a small, but

steady demand for sharkskin for many hundreds of years. It was originally valued for its abrasive properties and was used for finishing timber – the denticles acting like grains on sandpaper – and for polishing marble. This product was called shagreen, and was simply pieces of dried skin. The Japanese are credited with being the first to use shagreen on sword handles, where it provided a good grip, even when covered with blood. It was also in Japan that the first sharkskin-covered objects were made in the seventeenth century.

The idea was quickly adopted in Europe – particularly in France – where spectacular use was made of shagreen for covering books, instrument cases and scabbards. Often the denticles were ground, polished and dyed, sometimes forming striking patterns. The best known craftsman in sharkskin was

Jean-Claude Galluchat who worked in Paris in the 1760s and 70s, during the reign of Louis XV. The techniques used by Galluchat went out of favour in the nineteenth century, and examples of this type of work are now rare, and are often not recognised for what they are.

Although the use of sharkskin for high-quality work rapidly declined, it did lead to experiments with less exotic shark leathers. Some skins were used with their denticles intact, but in most cases the hides were too stiff. Techniques were developed for grinding away the denticles, but this was a delicate process, and not always satisfactory. It was not until about 1920 that the first successful chemical process was developed for removing denticles, while leaving the skin undamaged.

Shark leather is still manufactured today (see box p 144) and products made from it are

Australian entertainer Bob Dyer
with a 567-kg (1250-lb) tiger shark, and
his wife Dolly with a 178-kg (393-lb)
mako at Watsons Bay, Sydney, in the
1950s. Both held world records for
sharks during their fishing careers.

THE SHARK'S REVENGE

Considering the thousands of sharks
that are caught each year it is surprising
that more of their would-be captors are
not injured by them. Only a few dozen
cases have been recorded in which
professional fishermen or amateur
anglers have received serious wounds
while handling sharks.

One shark, however, did manage to
exact revenge on its tormentors in the
most unusual circumstances. In July
1970 a fisherman aboard a 13-m (43-ft)
powerboat off the Virginia coast hooked
a small, 2.3-kg (5-lb) shark and dragged
it aboard. Apparently bored or curious,
the fisherman stuffed a small bomb (of
the type used to train soldiers) down the
shark's throat, set the fuse, and then
threw the animal overboard again. As
those aboard waited, anticipating the
results of their little joke, the shark
doubled back and, with immaculate
timing, blew up exactly underneath the
boat. The boat promptly sank – a large
hole blown in its bottom planks – and
the despondent owner eventually faced
a bill of $5000 to pay for salvage and
extensive repairs.

Big game fisherman Alf Dean
(foreground) with five great white
sharks caught off the South Australian
coast. Some of Dean's records of the
1950s and 60s still stand today.

reputed to have twice the life of those made from conventional leather. Shark leather shoes are particularly hard-wearing, although the fact that the leather is extremely dense apparently prevents moisture from escaping through it, and this can be uncomfortable for the wearer.

A few other shark by-products are exploited commercially, although none is yet the basis of a major industry. A small trade has grown up around shark curios, such as mounted jaws and shark tooth jewellery – especially any made from the teeth of a large great white shark – which can command high prices. It appears, however, that there may be more potential in the medical use of shark products.

Various parts of sharks have been used in medicine by many cultures, and from the earliest times. Monsieur Pomet, the French

druggist, reported in 1730 that: 'The petrified teeth of this fish [*Canis Carcharias*] are what are called Glossopetrae, they are hung by the good woman about children's necks, in the imagination that they assist them in the time of cutting their teeth. They are also said to be a cordial, alexipharmick [antidote], and sudorifick [agent capable of causing sweating] taken inwardly, but I believe few have tried them.' In 1826 W. Ainslie wrote in his book *Materia Indica* that: 'The flesh of the Shark-Fish is supposed by the Hindoo medical writers to have peculiar virtues in several diseases; and is particularly noticed ... as a diet to be had recourse to in rheumatic affections.' Modern uses for shark products are less speculative. Shark corneas have been successfully transplanted into human eyes, and a synthetic skin for burn victims has been

manufactured using a chemical extracted from cartilage.

Sharks are among several species of gamefish, including marlin and tuna, that are sought all over the world. Their great size and fearsome reputation has made them attractive prizes for anglers.

The International Game Fishing Association recognises nine species of sharks – the blue, hammerhead, mako, porbeagle, school (tope), thresher, tiger, whaler and great white – which can bring anglers points in a competition. Strict rules govern fishing competitions and records are established by anglers for catching the heaviest shark on various standard weights of line.

Although big sharks, such as great whites and tigers, are the most spectacular sights at the weighing station, it is the mako (nicknamed 'blue dynamite') that is

Exciting moments during a game fishing contest off Port Stevens, on the east coast of Australia. A large great white shark, coaxed to the surface by blood and bait in the water, circles ominously before snatching a 2.1-m (7-ft), 80-kg (176-lb) dead whaler shark hanging from the stern of the boat (left). Seconds later it rises to the surface, its mouth gaping, showing the dead whaler inside (right). After 2½ hours the great white was hooked, but the hook pulled out after the shark had been played for 4½ hours on 37-kg (80-lb) line. Those on board estimated its weight at between 1800 and 2300 kg (4000 and 5000 lb).

most prized as a fighting fish. A mako will jump metres out of the water in an effort to dislodge the hook in its mouth, and can take many hours to conquer and capture. The great white has also been known to leap from the water when hooked, but most other species dive deep, and are gradually worn down by skill and persistence.

THE SHARKS WE EAT

Shark flesh is eaten all over the world, although it is not always identified as such because of public resistance to sharks as food. In the United States mako flesh is often sold as swordfish, which is a delicacy, and in Britain many spiny dogfish end up in fish and chip shops.

Apart from a public distaste for the idea of eating sharks, the only problem with marketing the flesh is that it can develop a taste and smell of ammonia if it is not handled properly. Sharks must be bled as soon as they are captured so that urea is removed from the flesh. This is usually done on board the fishing boat, and involves removing the head and belly flaps from the fish. After that has been done it is almost impossible for a layman to identify what remains as part of a shark.

Melbourne is the major outlet for sharks in Australia, although bins of school sharks (most of which are only about 40 cm [16 in] long), wobbegongs, angel sharks, and perhaps a few hammerheads and whalers are offered for sale at the Sydney fish markets most mornings. The majority of these end up in suburban fish shop windows, labelled as 'boneless fillets'.

A bin of small school sharks for sale at the Sydney fish markets.

Tab. VIII

Catulus major 1 großer hund fisch

Catulus 2 Canicula Ariftot. Kleiner Meer hund

Canis Marinus 3

Canicula 4 Galeus canis ein art der kleinen Meer hund

Muftelus Spinax 5 Stachd hund

Muftelus Læuis 6 glatt Seehund

Galeus 7 Hund fisch

Galeus Asterias feu stellatus Sternhund
8

9 Galeus Stellaris
Galeus Glaucus Venetanus Gesn.

Part Three

3

Facts About Sharks and Shark Attack

All 344 known
species of sharks
are described in detail
on the following pages. This
section also includes a
comprehensive guide to
shark records, and maps
showing the locations
of all documented
attacks on humans.

Illustrations of sharks
from Conrad Gesner's
Fisch-Buch, first
published in 1670.

SHARKS
OF THE WORLD

On the following pages are detailed all 344 known species of sharks, with illustrations of some of the most interesting and remarkable of them. The way in which sharks are classified into orders, families, genera and species – and the features that help in identifying them – are explained below.

The bull shark, the Ganges shark, the Zambezi shark, the river shark, the Nicaragua shark, shovelnose, the cub shark – these are all the same creature that is found and dreaded worldwide. Common names for sharks can vary from region to region. Abundant sharks or man-killers are dubbed with many names; rare kinds may not get any common name at all.

An international system of naming animals had become necessary by the beginning of the eighteenth century, since by then huge numbers of animals were known. The system of binominal nomenclature (two-name naming) devised by the Swedish naturalist Carl Linnaeus (1707-78) is still used today. Each animal has a generic or

genus name (a noun, given a capital letter), followed by a specific or species name (an objective). The names usually come from Greek or Latin, are often in compound forms, and are written in italics. For example, *Ginglymostoma brevicaudatum* comes from the Greek for hinge-mouth, and the Latin for short-tailed. Sometimes names commemorate people or places. Scientific books and journals add the name of the person who first described and named the species, with the date, so the Port Jackson shark is given as *Heterodontus portusjacksoni* (Meyer, 1793).

The system goes further. The base units of the system are the species. (Broadly speaking, if two animals can produce fertile

offspring, they are of the same species.) Related species are grouped in genera; genera in families; families in orders; orders in classes; classes in phyla; and phyla in the animal kingdom.

The animal kingdom consists of about 30 phyla, each grouping animals with a particular body plan. For instance, spiders and people are in different phyla, but sharks and people are in the same phylum because they both have backbones. The backboned animals (phylum Chordata) are divided into classes that include reptiles, birds, mammals and the three classes of fishes: jawless fishes, those with bony skeletons, and those with a rod of cartilage for a backbone (the class which includes sharks).

The class of cartilaginous fishes – Chondrichthyes – consists of sharks, rays, chimaeras and ratfishes. Sharks and rays have a unique sort of blood, and have upper jaws slung from their skulls, not fixed as with chimaeras and

---

HOW THE GUIDE TO SHARK SPECIES IS ORGANISED

The charts that start on page 156 list all known sharks by their most recent scientific names – others that may be found in modern books appear in the index (p 202). Space allows only the more widely used common names to be included, and again these are all indexed. Each shark's range is given with a series of numbers that relate to the map (right). It divides the world into regions which are for convenience only – they are not based on water temperatures or currents. Again, lack of space prevents the inclusion (where known) of a full description of the preferred habitat of each shark, such as rocky or sandy bottom. The first size given is the maximum known; the following two are the average for females (♀) and males (♂).

Frequent new finds and current research make any list of sharks obsolete shortly after publication. This list was assembled with considerable help from the two-volume *Sharks of the World* which was compiled by Dr Leonard J.V. Compagno for the United Nations Food and Agriculture Organisation and published in 1984. It gives considerably more information about each shark than it has been possible to include here, but much still remains to be discovered about many of the world's rarer species.

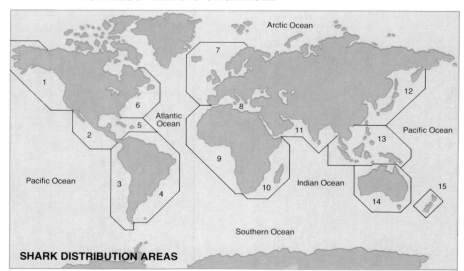

SHARK DISTRIBUTION AREAS

It is obviously difficult to be precise about the global distribution of animals that move freely in the oceans. The data given in the charts is approximate only, and indicates that particular species have been found in the region(s) indicated, but not necessarily that they never occur in those that are not

marked. In some cases the known range of a species might be much more localised than is shown here, but it has been impossible to be more precise in the limited space available. The extent to which some species, such as the bull shark, venture into fresh water has not been mapped.

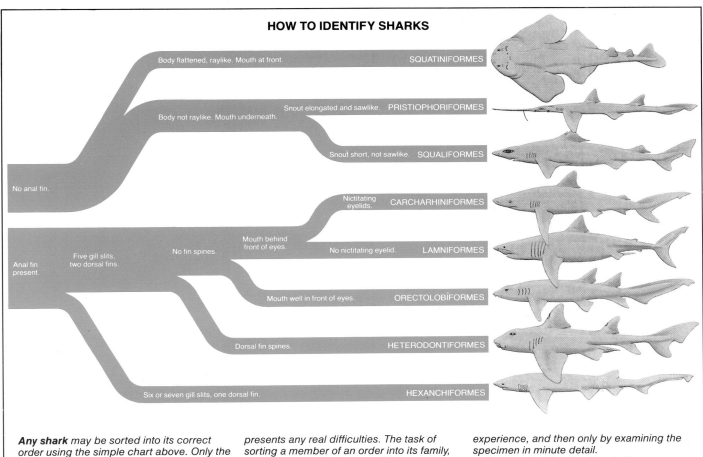

HOW TO IDENTIFY SHARKS

Body flattened, raylike. Mouth at front. — SQUATINIFORMES

Body not raylike. Mouth underneath. — Snout elongated and sawlike. PRISTIOPHORIFORMES

Snout short, not sawlike. SQUALIFORMES

No anal fin.

Nictitating eyelids. CARCHARHINIFORMES

Mouth behind front of eyes. — No nictitating eyelid. LAMNIFORMES

Anal fin present. — Five gill slits, two dorsal fins. — No fin spines. — Mouth well in front of eyes. ORECTOLOBIFORMES

Dorsal fin spines. HETERODONTIFORMES

Six or seven gill slits, one dorsal fin. HEXANCHIFORMES

Any shark may be sorted into its correct order using the simple chart above. Only the task of examining the eye to see if it has a nictitating membrane, in order to distinguish between carcharhinifomes and lamniformes, presents any real difficulties. The task of sorting a member of an order into its family, genus and species is more difficult, however. In many cases this can only be done accurately by an expert after many years of experience, and then only by examining the specimen in minute detail.

Illustrated regional guides to the more common shark species are available for some parts of the world.

---

ratfishes, so they are placed in a sub-class, Plagiostomi (the oblique-mouths). Sharks differ from rays because their pectoral fins are separate from the sides of their heads; they have free upper eyelids, and gill openings on their heads (in rays they are located below the pectoral fins), so sharks are put in a super-order, the Selachimorpha.

The Selachimorpha comprises eight orders of sharks. HEXANCHIFORMES (six-gilled shapes), are the frilled and cow sharks. SQUALIFORMES (the Latin word *squalus* originally encompassed all sharks), are the dogfish sharks that include the smallest shark – the spined pygmy shark – and the world's third-largest

fish – the Greenland shark. PRISTIOPHORIFORMES (saw-carriers) are the sawsharks, and SQUATINIFORMES (angel-fish shapes) are the angel sharks. HETERODONTIFORMES (different teeth, for their small front teeth and large back teeth) are the bullhead sharks. ORECTOLOBIFORMES (extended [tail]-lobes), the carpet sharks – which includes the whale shark and wobbegongs. LAMNIFORMES (from the Greek lamna, a fish of prey), are the mackerel sharks that include the goblin, megamouth, basking, great white and thresher sharks. CARCHARHINIFORMES (jagged rasp, because of teeth or skin, or both), the ground sharks, include the

requiem sharks and hammerheads.

As the species *Homo sapiens* (Linnaeus, 1758) increases in numbers, and exploits the oceans more and more for food, new species of sharks will be discovered. Other species will be added to the lists just by scientists reappraising the ones already known. Each species has, or once had, a holotype or type specimen, kept in alcohol, or formaldehyde, glycerine and seawater, in a museum or other institution. To reclassify a species, a zoologist may review the published description and draw different conclusions, or re-examine the type specimen. Scientists often disagree on links between species, or on species themselves.

Scientific name	Common names	Distribution*	Size (cm)*	Colour	Notes

HEXANCHIFORMES (Frilled and cow sharks)
Chlamydoselachidae (Frilled sharks)

Scientific name	Common names	Distribution*	Size (cm)*	Colour	Notes
Chlamydoselachus anguineus	Frilled shark, frilled-gilled s, frill s	1,3,4,7,9,10,12,14,15. Bottom, deep	196	Mouse-grey or dusky, paler below	Name means mantled shark + eel; a rare, primitive shark resembling a snake. It has six gill slits, the first nearly encircling the head and looking like a frilled collar.

Hexanchidae (Cow sharks, sixgill sharks, sevengill sharks)

Scientific name	Common names	Distribution*	Size (cm)*	Colour	Notes
Heptranchias perlo	Sharpnose seven-gill shark, perlon s, sevengill s	4,5,6,8,9,10,11,12,13,14,15. Bottom, mainly deep, sometimes shallow inshore	137	Mouse-grey, off-white below	Fierce when caught, but small size prevents it from being considered dangerous.
Hexanchus griseus	Bluntnose sixgill shark, cow s, grey s, mud s, bulldog s	1,2,3,4,5,6,7,8,9,10,12, 13,14,15,16. All levels, mainly deep	482	Mouse-grey, charcoal or russet	Taken for food, but its liver is toxic. Litters range from 20 to 108; size at birth about 65 cm (26 in). Has been found at depth of over 1800 m (5900 ft). IGFA game fish.
Hexanchus vitulus	Bigeye sixgill shark, lesser six-gill s, calf s	5,9,10,12,13. Warm temperate and tropical seas, generally bottom	180	Mouse-grey, off-white below	
Notorynchus cepedianus	Broadnose seven-gill shark, ground s, cow s, broad snout	1,3,4,9,10,12,14,15. Often close to and on surface. Temperate seas	290 ♀240 ♂188	Sandy-grey or russet with dark spots	Aggressive, fights when caught. In captivity has attacked people. Suspected of unprovoked attacks. Possibly prefers rays and other sharks as food.

SQUALIFORMES (Dogfish sharks)
Echinorhinidae (Bramble sharks)

Scientific name	Common names	Distribution*	Size (cm)*	Colour	Notes
Echinorhinus brucus	Bramble shark, spiny s, alligator dogfish, prickle s	4,6,7,8,9,10,11,12,14,15. Bottom, deep, occasionally shallow	310 ♀220 ♂162	Purple-grey or brown, often spotted black	Fat, sluggish, usually covered with evil smelling mucus. In South Africa its liver oil is highly prized for medicine. Is covered in thorny denticles.
Echinorhinus cookei	Prickly shark, Cooke's s, spiny s	1,2,3,12,15,16. Bottom, shallow and deep	400	Grey to brown, white under-side to snout	Has much smaller prickles than the bramble shark.

Squalidae (Dogfish sharks)

Scientific name	Common names	Distribution*	Size (cm)*	Colour	Notes
Aculeola nigra	Hooktooth dogfish	3. Bottom, deep	60 ♀56	Charcoal to black, fin tips white	The 60 rows of teeth in each jaw are slightly hooked.
Centrophorus acus	Needle dogfish	5,12. Deep	81	Light to dark grey	Like all members of the genus Centrophorus, has huge green eyes.
Centrophorus granulosus	Gulper shark, gulper	5,7,8,9,10. Bottom, deep	150	Sandy, paler below	Exploited in Japan for the oil in its large liver.
Centrophorus harrisoni	Dumb gulper shark	14. Deep	80	Colour not known	So far, known positively only from southeast mainland Australia.
Centrophorus lusitanicus	Lowfin gulper shark	7,9,10,12. Deep	160 ♀116 ♂100	Dusky, fin edges white or translucent	Name refers to its occurrence off Portugal, which was part of the Roman province of Lusitania.
Centrophorus moluccensis	Smallfin gulper shark, Endeavour dogfish	10,12,13,14,16. Bottom, deep	98 ♀93 ♂77	Grey, white below	Plentiful off Mozambique and South Africa.
Centrophorus niaukang	Taiwan gulper shark	12. Deep	154	Colour not known	Only known off Taiwan where its liver oil is used.
Centrophorus squamosus	Leafscale gulper shark, Nilson's deepsea dogfish	7,9,10,12,13,14,15. Deep, also pelagic	158 ♀147	Charcoal	Has overlapping denticles on small stalks covering its flanks.

Frilled shark
Chlamydoselachus anguineus
196 cm

Sharpnose sevengill shark
Heptranchias perlo
137 cm

Bramble shark
Echinorhinus brucus
310 cm

*For note on size, and distribution map, see box p154.

Scientific name	Common names	Distribution*	Size (cm)*	Colour	Notes
Centrophorus tessellatus	Mosaic gulper shark, tessellated deepwater s	12,16. Bottom, deep	89	Fawn, pale white bands at fin edges	Centro-phorus means spine-bearing. All in this genus have a strong spine on each dorsal fin.
Centrophorus uyato	Little gulper shark, southern dogfish	5,8,9,10. Bottom, deep	100 ♀82 ♂88	Light mouse-grey, paler below	
Centroscyllium fabricii	Black dogfish	6,7,9. Baffin Island and Greenland. Bottom, deep, sometimes surface	84 ♀64	Chocolate brown, white fin spines	Occurs off the west coast of Africa, also close to Arctic areas where it is commonly taken from near freezing waters.
Centroscyllium granulatum	Granular dogfish	Falkland Islands. Deep	28	Dusky to black	All members of this genus are luminescent.
Centroscyllium kamoharai	Bareskin dogfish	12. Deep	44 ♀43	Charcoal	The denticles are few and far apart.
Centroscyllium nigrum	Combtooth dogfish, Pacific black dogfish, black s	2,3,16. Bottom, deep	50 ♀50 ♂37	Stippled black, white tips to fins	All members of this genus have several rows of comb-like teeth.
Centroscyllium ornatum	Ornate dogfish	11. Bottom, deep	30	Black	Centro-scyllium means spiny small shark. They all have fine dorsal spines, the second much longer than the first.
Centroscyllium ritteri	Whitefin dogfish	12. Deep	43 ♀42	Mouse-grey, black below, white edged fins	
Centroscymnus coelolepis	Portuguese dogfish, Portuguese s	6,7,8,9,12. Bottom, deep	114 ♀92 ♂92	Chocolate brown	Used to be taken in large numbers off Portugal. Netted at 2640 m (8922 ft), deepest for any shark.
Centroscymnus crepidater	Longnose velvet dogfish, golden dogfish	3,7,9,10,11,14,15. Bottom, deep	90	Dusky black, golden or dark brown	
Centroscymnus cryptacanthus	Shortnose velvet dogfish	4,9. Bottom, deep	104 ♀103 ♂78	Dusky or charcoal	Crypt-acanthus means hidden spine. Dorsal fin spines are hidden within the fins.
Centroscymnus macracanthus	Largespine velvet dogfish	Straits of Magellan. Deep	68	Dusky	Only one specimen known, from Straits of Magellan, described 1906.
Centroscymnus owstoni	Roughskin dogfish, Owston's spiny dogfish	5,12,14,15. Bottom, deep	78	Dusky	Named after Alan Owston, an American and a collector of Japanese fish who lived in Yokohama.
Centroscymnus plunketi	Plunket shark, Lord Plunket's s	14,15. Bottom, deep	170 ♀150 ♂115	Chocolate brown	Named after a governor of New Zealand. Litters up to 36. Males and females of same size form separate, large schools.
Cirrhigaleus barbifer	Mandarin dogfish	12,13,14,15. Bottom, deep	122 ♀107 ♂86	Mouse-grey, pale below	Has mandarin-moustache-like barbels or flaps to nose, possibly trailed on sea-floor to detect prey.
Dalatias licha	Kitefin shark, Bonnaterre's deepwater s, seal s, black s	5,6,7,8,9,10,11,12,14,15,16. Tropical and warm temperate seas; deep	159 ♀138 ♂99	Chocolate, often black-spotted	Skin is used for an abrasive leather. Has fringed lips, very strong jaws, and smooth, thorn-like upper teeth, while lower teeth are serrated, wide, triangular. Sluggish but takes swifter prey such as bonito, perhaps by ambush. No one is sure why males more often have fuller stomachs than females.
Deania calcea	Birdbeak dogfish, shovelnosed s, Thompson's s	3,7,9,10,12,14,15. Bottom, deep	111 ♀90 ♂80	Pale mouse-grey	All members of this genus have teeth that develop differently in males and females.
Deania histricosa	Rough longnose dogfish	9,12. Bottom, deep	109 ♀107 ♂84	Dusky	So far known only off Madeira and Japan.
Deania profundorum	Arrowhead dogfish	6,9,100,13. Bottom, deep	76 ♀73 ♂55	Dusky	Little known. Has a long, pointed snout.

Leafscale gulper shark
Centrophorus squamosus
158 cm

Mandarin dogfish
Cirrhigaleus barbifer
122 cm

Arrowhead dogfish
Deania profundorum
76 cm

Scientific name	Common names	Distribution*	Size (cm)*	Colour	Notes
Deania quadrispinosum	Longsnout dogfish	9,10,14. Deep	114	Chocolate	Little known. In Australia, southeastern and southern mainland coast only.
Etmopterus baxteri	New Zealand lanternshark	15. Bottom, deep	75 ♀75 ♂66	Dusky with large black spots below	The skin of all lanternsharks secretes a luminous mucus.
Etmopterus brachyurus	Shorttail lanternshark	13. Bottom, deep	24	Brown, black fin stripes, black below	
Etmopterus bullisi	Lined lanternshark	5,6. Bottom, deep		Charcoal, yellowish spot on head	Little known. No length can be given because the largest measured specimen was immature.
Etmopterus decacuspidatus	Combtoothed lanternshark	13. Bottom, deep	29	Brown, black fin stripes, black below	Known from only one specimen described in 1966. Deca-cuspidatus = ten-pointed. Upper teeth have 8-10 cusplets each side.
Etmopterus gracilispinis	Broadbanded lanternshark	4,6,9. Bottom, deep, and at moderate depths over very deep water	33 ♀33 ♂25	Charcoal, broad black fin stripes	
Etmopterus granulosus	Southern lanternshark	3,4,9. Deep	40	Brown, black below, black fin streaks	Lengths of mature males and females not known; an immature male was 32 cm (13 in.).
Etmopterus hillianus	Caribbean lanternshark, blackbelly dogfish	5,6. Bottom, deep	50 ♀30 ♂26	Chocolate, yellow spot on head	
Etmopterus lucifer	Blackbelly lanternshark, Moller's deepsea dogfish	4,10,12,13,14,15. Bottom, deep	42 ♀34 ♂35	Dusky, paler fins, black fin stripes	Has survived in captivity in Japan.
Etmopterus polli	African lanternshark	9. Bottom, deep	24 ♀24 ♂23	Charcoal, dark bands above fins and tail	
Etmopterus princeps	Great lanternshark	6,7,9. Bottom, deep	75 ♀55	Dusky, black below	Has been found more than 2000 m (6500 ft) down.
Etmopterus pusillus	Smooth lanternshark	4,5,7,9,10,12. Bottom, deep, and surface to deep in oceanic waters	75 ♀42 ♂35	Black, yellow spot on head, pale fin edges	
Etmopterus schultzi	Fringefin lanternshark	5. Bottom, deep	30 ♀29 ♂27	Fawn with dark marks, dark below	All fins have wide fringes of horny 'hair'.
Etmopterus sentosus	Thorny lanternshark	10. Bottom, deep		Charcoal with black marks, darker below	Sentosus means thorny. This shark has two lines of large spiny denticles on flank. An immature specimen was 27 cm (11 in) long.
Etmopterus spinax	Velvet belly	7,8,9. Bottom, medium-deep and deep	60 ♀34 ♂34	Brown, black below with greenish line	Forms large schools; common.
Etmopterus unicolor	Brown lanternshark	12	53	Dusky	
Etmopterus villosus	Hawaiian lanternshark	16. Bottom, deep	46 +	Dusky with black marks, darker below	
Etmopterus virens	Green lanternshark, green dogfish	5. Deep	23 ♀23 ♂21	Chocolate, black-green sheen below	Catching large prey may be done in groups.
Euprotomicroides zantedeschia	Taillight shark	4,9. Oceanic waters off Uruguay and Cape Province, South Africa	42	Brown, black below, light fin edges	Among the smallest of all sharks. A gland below its tail secretes a luminous blue substance. Known only from two specimens (Uruguay, 1966, and South Africa, 1980).

African lanternshark
Etmopterus polli
24 cm

Taillight shark
Euprotomicroides zantedeschia
42 cm

Pygmy shark
Euprotomicrus bispinatus
27 cm

*For note on size, and distribution map, see box p154.

Scientific name	Common names	Distribution*	Size (cm)*	Colour	Notes
Euprotomicrus bispinatus	Pygmy shark, slime shark	Surface to midwaters of Indian, Pacific and South Atlantic Oceans	27 ♀24 ♂19	Fawn to charcoal, clear fin edges	Found near surface at night, making big vertical journeys by day (up to 1500 m, 4900 ft, each way). Luminescent undersides.
Heteroscymnoides marleyi	Longnose pygmy shark	9,10. Probably oceanic	28	Pale brown, translucent edges to fins	Known only from two specimens (Natal, Africa, 1934 and near Ascension Island in the Atlantic, 1980).
Isistius brasiliensis	Cookiecutter shark, cigar s, luminous s, Brazilian s	4,5,9,12,14,16. Oceanic, surface to deep waters	50 ♀44 ♂35	Chocolate, green pupils to large eyes	Isistius from Isis, goddess of light, because it emits a vivid green light from its belly. The glow possibly attracts predators which then fall prey to this shark. It has very strong jaws and long teeth; can clamp itself to prey as big as itself by its lips; it bites, twists and gouges out a plug of flesh. Victims include dolphins and whales. Has been known to attack rubber casings to sonar domes of submarines. Swallows loose lower teeth, perhaps for calcium. Possibly makes vertical journeys of more than 2 or 3 km (well over 1 mile) up and down each day.
Isistius plutodus	Largetooth cookiecutter shark	5,12. Moderate and possibly deep waters	42	Chocolate, paler patch under head	Plutodus = plenty-toothed. Its huge lower jaw teeth are, for its size, the largest of any living shark. Can cut very deep plugs of flesh from victims. A plug of flesh in the stomach of one was the diameter of its mouth but twice as long.
Scymnodalatias sherwoodi	Sherwood dogfish	15. Possibly oceanic	80	Dark brown	Known from only one specimen washed up on a beach in the 1920s in Canterbury, New Zealand.
Scymnodon obscurus	Smallmouth velvet dogfish	4,5,7,9. All levels of oceanic waters	59 ♀59 ♂51	Black	
Scymnodon ringens	Knifetooth dogfish	7. Bottom, deep	110	Colour not known	Has enormous, knife-edged teeth in lower jaw.
Scymnodon squamulosus	Velvet dogfish	12. Deep	69 ♂49	Black	
Somniosus microcephalus	Greenland shark, gurry s, sleeper s	6,7,9, also Greenland, White Sea, Kerguelen I. Surface in cold waters	640 ♀373 ♂293	Greyish-brown	In Arctic and North Atlantic waters, large numbers gather to feed around fishing and sealing operations. When gorging they are apparently oblivious to blows from clubs and other weapons. Their size alone was probably the basis for stories of attacks on people in boats; is now believed harmless. Is taken almost without resistance – simply being lured to surface then lifted with gaffs. Easily fished through holes in ice. Inuit (Eskimos) traditionally used skin for boots, and formed small knives for cutting hair from lower teeth. Usually has one parasitic crustacean attached to each eye which may attract prey towards its host. This shark eats mammals, alive or dead; an entire reindeer minus antlers was found inside one. IGFA game fish.
Somniosus pacificus	Pacific sleeper shark, sleeper s, North Pacific s	1,2,12, also Siberian coast. Shallow in cold waters, elsewhere deep	700 ♀400	Slate	As name implies, sleeper sharks are sluggish yet they can catch swift-moving prey, including seals. Flesh is toxic unless dried or semi-putrid; symptoms of poisoning are like drunkenness.

Cookiecutter shark
Isistius brasiliensis
50 cm

Velvet dogfish
Scymnodon squamulosus
69 cm

Greenland shark
Somniosus microcephalus
640 cm

Scientific name	Common names	Distribution*	Size (cm)*	Colour	Notes
Somniosus rostratus	Little sleeper shark	7,8,12. Bottom, moderate and deep waters	140 ♀108 ♂71		Its comparatively small size shows that it is impossible to generalise about any genus.
Squaliolus laticaudus	Spined pygmy shark	4,7,9,10,12,13. Offshore and oceanic; tropical waters, deep	25 ♀21 ♂18	Mouse-grey, grey or black	Smallest of all sharks. Unique in having a spine on first dorsal fin only. Underside is luminescent. This eliminates any shadow when light comes from above, and predators are less likely to see it.
Squalus acanthias	Piked dogfish, spiny dogfish, skittledog, white-spotted dogfish, codshark, thorndog	Worldwide in temperate and subarctic waters. All levels	160 ♀97 ♂79	Slate, few white spots, paler below	Probably the most common shark. At the beginning of the century 27 million were taken off the coast of Massachusetts every year. Flesh is prized by Italians, eaten in Britain as 'rock salmon' and 'flake' in Australia. Its abundance and non-specialised anatomy mean it is often studied in laboratories, and probably more is known of this than any other species. Gestation period is 18 to 24 months, longer than elephants or whales. May live over 30 years; some scientists estimate nearly 100. Is hated by fishermen who fear teeth and mildly poisonous spines on dorsal fins, and dread the damage it does to nets and catches, but is an important food fish. Is also used for liver oil, pet food, fishmeal, hide, fertiliser. Will enter brackish water. Forms huge schools, sometimes segregated by sex and size. Migrates to stay in water between 7° and 15°C (45° to 59°F); one tagged off west United States coast was found 7 years later off Japan, 6500 km (4000 miles) away. IGFA game fish.
Squalus asper	Roughskin spurdog, roughskin spiny dogfish	5,10,16. Bottom, deep	118 ♀103 ♂87	Chocolate, paler below	Has litters of about 21, each 26 cm (10 in) long.
Squalus blainvillei	Longnose spurdog, Blainville's dogfish	7,8,9,12. Bottom, deep and shallow	95 ♀60 ♂50	Mouse-grey, white marks, white below	Common; often forms large schools.
Squalus cubensis	Cuban dogfish	4,5,6. Bottom, moderate and deep	110 ♀62 ♂62	Mouse-grey, paler below	Common; forms large schools. A large parasitic crustacean lives in its mouth.
Squalus japonicus	Japanese spurdog	12,13. Bottom, deep	91 ♀79	Mouse-grey, white edged dorsal fins	
Squalus megalops	Shortnose spurdog, Spiky Jack, piked dogfish	9,10,12,13,14. Moderate to deep waters	71 ♀62	Dusky to grey, paler below	Megal-ops = large eyes. Like *Squalus acanthias,* this shark has a two-year gestation period (longer than whales).
Squalus melanurus	Blacktailed spurdog	16. Deep	75	Brown, black edges to tail and fins	Specimens found only off New Caledonia. Lashes about when caught and has a dangerous spine on its second dorsal fin.
Squalus mitsukurii	Shortspine spurdog	3,4,5,6,10,11,12,13, 14,15,16. All levels	110 ♀72 ♂77	Dusky, white edges to fins	Common. Sometimes found off New Zealand in water only 4 m (12 ft) deep.
Squalus rancureli	Cyrano spurdog	16. Deep	77 ♀71	Brown	Named for its very long pointed snout. So far, specimens found off Vanuatu only.

Oxynotidae (Roughsharks)

Scientific name	Common names	Distribution*	Size (cm)*	Colour	Notes
Oxynotus bruniensis	Prickly dogfish, humantin	14,15. Bottom, medium and deep waters	72 ♀72 ♂60	Mouse-grey	All members of this single-genus family have large, sail-like, high dorsal fins above a stout body.

Spined pygmy shark
Squaliolus laticaudus
25 cm

Piked dogfish
Squalus acanthias
160 cm

Prickly dogfish
Oxynotus bruniensis
72 cm

*For note on size, and distribution map, see box p154.

Scientific name	Common names	Distribution*	Size (cm)*	Colour	Notes
Oxynotus caribbaeus	Caribbean roughshark	4. Bottom, deep	49	Brown to grey, with dark marks	All roughsharks have upper teeth that are larger towards the back, and lower teeth that point backwards.
Oxynotus centrina	Angular roughshark	7,8,9. Bottom, medium and deep waters	150	Mouse-grey, darker on head	Oxy-notus = sharp back.
Oxynotus paradoxus	Sailfin roughshark, sharp-back shark	7,9. Bottom, deep	118	Dusky, paler below	Roughsharks' large bodies and oily livers probably make them able to hover and coast at the bottom to seek food.

PRISTIOPHORIFORMES (Sawsharks)
Pristiophoridae (Sawsharks)

Scientific name	Common names	Distribution*	Size (cm)*	Colour	Notes
Pliotrema warreni	Sixgill sawshark	10. Bottom, medium to deep waters	136 ♀123 ♂97	Pale grey	Plio-trema = plentiful openings (gills). Sawsharks' 'teeth' on their snouts are enlarged denticles.
Pristiophorus cirratus	Longnose sawshark, common sawshark, little sawshark	14. Inshore, bays and estuaries, and offshore in medium to deep waters	137	Sandy or greyish, brown marks	Sawsharks differ from sawfish in having two barbels near the saw-like snout, and side gills. Much exploited for food.
Pristiophorus japonicus	Japanese sawshark	12. Bottom, offshore	136	Olive-brown	Sawsharks' long barbels probably detect prey disturbed by the snout.
Pristiophorus nudipinnis	Shortnose sawshark, southern sawshark	14. Bottom, medium to deep waters	122	Mouse-grey, paler below	The 'teeth' on the saws of this family are extremely sharp, but, unlike sawfishes, these sharks are inoffensive.
Pristiophorus schroederi	Bahamas sawshark, American sawshark	5,6. Bottom, deep, tropical waters	80	Pale mouse-grey, white below	Sawsharks bear live young. Until birth their saw teeth are folded back so as not to injure the mother.

SQUATINIFORMES (Angelsharks)
Squatinidae (Angelsharks, sand devils, monk sharks)

Scientific name	Common names	Distribution*	Size (cm)*	Colour	Notes
Squatina aculeata	Sawback angel-shark, monkfish	8,9. Bottom, medium to deep waters	188	Sandy, with symmetrical white spots	Angelsharks resemble rays but, unlike rays, fins are not attached to the head. This species has large spines on snout.
Squatina africana	African angel-shark	10. Bottom, surf zone and deep waters	108 ♀99 ♂78	Mouse-grey, white spots, white below	Like all angelsharks or sand devils, often buries itself in sand or mud, with only eyes and top of body showing.
Squatina argentina	Argentine angel-shark	4. Bottom	170	Colour not known	All angelsharks can, if provoked, inflict severe cuts, having strong jaws and sharp teeth.
Squatina australis	Australian angel-shark, monkfish	14. Bottom, shallow inshore to deep waters	152	Sandy, many white and grey spots	The large pectoral fins of angelsharks can be cut into steaks.
Squatina californica	Pacific angelshark	1,2,3. Bottom, inshore, shallow and medium to deep waters	152 ♂95	Speckled sandy/russet, white below	Angelsharks' camouflage colouring allows them to lie in ambush waiting for prey. Can bite swiftly with extremely sharp teeth, and people should be very wary of this shark.
Squatina dumeril	Sand devil, Atlantic angelshark	4,5,6. Bottom, shallow inshore to deep waters	152 ♂100	Russet to light grey, white below	Called sand devil because it can inflict severe cuts on fishermen who take it.
Squatina formosa	Taiwan angelshark	12. Bottom, deep		Colour not known	Formosa = pretty, but name refers to Taiwan, not this member of a weird-looking family. Known from only one immature specimen (1972).

Caribbean roughshark
Oxynotus caribbaeus
49 cm

Sixgill sawshark
Pliotrema warreni
136 cm

Australian angelshark
Squatina australis
152 cm

Scientific name	Common names	Distribution*	Size (cm)*	Colour	Notes
Squatina japonica	Japanese angel-shark	12. Bottom	200	Colour not known	In China, much exploited for food and for hide to make shagreen.
Squatina nebulosa	Clouded angel-shark	12. Bottom	163	Colour not known	This, like several angel sharks, has survived in captivity.
Squatina oculata	Smoothback angelshark, monkfish	8,9. Medium to deep waters	160	Sandy/grey, black and white spots	Has large spines on the snout and head but not on the back, hence the name 'smoothback'.
Squatina squatina	Angelshark, monkfish, angel-fish	7,8,9. Bottom, close inshore to moderately deep waters	183 ♀146	Sandy-grey/ greenish, white below	Esteemed for food in Mediterranean countries (occasionally, under a sauce, masquerading as lobster). Occurs off Sweden and was named and described by Linnaeus in 1758.
Squatina tergocellata	Ornate angelshark	14. Bottom, medium to deep waters	55	Tawny, prominent rings, blue spots	Has fringed nostrils (present to a lesser degree in all angelsharks). Spines on snout and head, small spines on back.
Squatina tergocellatoides	Ocellated angel-shark	12	63	Prominent rings on fins only	Known from only one specimen taken near Taiwan in the early 1960s.

HETERODONTIFORMES (Bullhead sharks)
Heterodontidae (Bullhead sharks, horn sharks)

Scientific name	Common names	Distribution*	Size (cm)*	Colour	Notes
Heterodontus francisci	Horn shark, horned shark	1,2. Bottom, moderately deep to very shallow waters	122 ♀58 ♂64	Sandy/grey, dark spots, yellow below	Most horn or bullhead sharks lay eggs in spiral flanged egg cases by which they get wedged in crevices. Dorsal fin spines are made into jewellery.
Heterodontus galeatus	Crested bullhead shark, crested Port Jackson s	14. Bottom, inshore and moderately deep waters	152	Tan, dark band below eyes	Egg case has tendrils up to 2 m (6 ft) long that anchor it to seaweeds. Young 17 cm (6.7 in) at birth.
Heterodontus japonicus	Japanese bullhead shark, horned s	12. Bottom, shallow and moderately deep waters	120	Russet-brown, saddle-like areas on back	Several females lay their eggs in one 'nest'. Like other bullheads, uses its broad, paddle-like fins to crawl on bottom.
Heterodontus mexicanus	Mexican hornshark	2,3. Bottom, shallow	70	Bronze-tan, black dots, head stripe	All bullhead sharks grind shellfish between their flat teeth. Some thrive and even breed in captivity.
Heterodontus portusjacksoni	Port Jackson shark, oystercrusher, pigfish, bulldog s	14,15. Bottom, shallow and moderately deep waters	165 ♀112 ♂87	Fawn/greyish, dark bars on flanks	A great deal is known about this shark from tagging and observation. Common, nocturnal. Has 'rest' areas, sometimes used by as many as 16 at a time. Ranges considerable distances from breeding areas, as much as 850 km (530 miles). Females have been seen carrying their egg cases, probably to wedge them in rock crevices. Harmless, although can bite.
Heterodontus quoyi	Galapagos bullhead shark	3. Bottom, inshore and moderately deep waters	59	Fawn/dusky, regular large black spots	Hetero-dontus = different teeth. Both jaws have pointed front teeth for grasping, and flat back teeth for crushing.
Heterodontus ramalheira	Whitespotted bullhead shark, Mozambique bullhead	10,11. Bottom, deep	83	Dark russet, white spots, cream below	Bullhead sharks are the oldest unchanged sharks; fossils have been found in rocks 200 million years old.
Heterodontus zebra	Zebra bullhead shark	12,13,14. Bottom, moderately deep waters	122	Dusky/black, zebra-like stripes	Bullhead sharks have blunt heads with large knobs above each eye.

Angelshark
Squatina squatina
183 cm

Mexican hornshark
Heterodontus mexicanus
70 cm

Port Jackson shark
Heterodontus portusjacksoni
165 cm

Scientific name	Common names	Distribution*	Size (cm)*	Colour	Notes
ORECTOLOBIFORMES (Carpetsharks)					
Parascyllidae (Collared carpetsharks)					
Cirrhoscyllium expolitum	Barbelthroat carpetshark	13. Bottom, offshore, moderately deep tropical waters	33	Fawn, dark blotches across back	Known from only one specimen taken in the China Sea (1913).
Cirrhoscyllium formanosum	Taiwan saddled carpetshark	12. Bottom, moderately deep waters	39 ♀37 ♂37	Pale, six saddle-like patches	Members of this family can change colour slightly to blend into the sea bed shades.
Cirrhoscyllium japonicum	Saddle carpetshark	12	49	Fawn with nine darker saddles	Cirrho-scyllium means orange-coloured dogfish.
Parascyllium collare	Collared carpetshark	14. Bottom, moderately deep waters	87 ♀86 ♂82	Tawny, dark collar and saddles	Occurs only around Tasmania and the southeast of mainland Australia.
Parascyllium ferrugineum	Rusty carpetshark, rusty catshark	14. Bottom, moderately deep waters	75	Mouse-grey, dark bands and spots	From southern Australian waters only.
Parascyllium multimaculatum	Tasmanian carpetshark, T. catshark, T. spotted cat s	14. Bottom, inshore	75 ♂73	Ash, many dark spots, white below	Found near river mouths and rocks in Tasmanian waters only.
Parascyllium variolatum	Necklace carpetshark, varied catshark	14. Bottom, moderate to deep waters	91	Fawn, dark collar, many white spots	Little known. Sometimes caught in rock lobster pots.
Brachaeluridae (Blind sharks)					
Brachaelurus waddi	Blind shark, brown catshark	14. Bottom, inshore very shallow and moderately deep waters	122	Fawn/dusky, white spots, yellow below	Called 'blind shark' because it closes its thick eyelids when taken out of the water.
Heteroscyllium colcloughi	Bluegrey carpetshark, Colclough's shark	14. Bottom, inshore	60	Grey, white below	Known only off Queensland.
Orectolobidae (Wobbegongs)					
Eucrossorhinus dasypogon	Tasselled wobbegong	13,14. Bottom, coral reefs and inshore waters	117	Fawn with dark network	Name means well-fringed nose and shaggy beard. The fringe and beard are formed of flaps of skin. Has reputation as a man-killer in Papua New Guinea, unsubstantiated.
Orectolobus japonicus	Japanese wobbegong	12,13. Bottom, inshore waters	103	Light and dark spots and lines	Wobbegongs have powerful jaws and dagger-like front teeth and can be dangerous if provoked or stepped on.
Orectolobus maculatus	Spotted wobbegong	12,13,14. Bottom, very shallow and moderately deep waters	320 ♀165 ♂165	Brown with lighter bars and blotches	Often seen climbing, half out of water, from one rock pool to another. Its attractive skin is used for bags/shoes. Wobbegongs possibly catch prey that has merely wandered near or even nibbled on their 'beards'. This shark is disliked by lobster fishers because it gets into pots to eat bait and lobsters.
Orectolobus ornatus	Ornate wobbegong, carpet s, banded wobbegong	12,13,14. Bottom, shallow inshore	288	Brown, light and dark marbling	The tough, patterned skin is used for leather.

Necklace carpetshark
Parascyllium variolatum
91 cm

Blind shark
Brachaelurus waddi
122 cm

Ornate wobbegong
Orectolobus ornatus
288 cm

*For note on size, and distribution map, see box p154.

Scientific name	Common names	Distribution*	Size (cm)*	Colour	Notes
Orectolobus wardi	Northern wobbegong	14. Bottom, inshore, tropical	45 +	Ochre, black fin dots, 3 spots on back	Wobbegongs do well in captivity.
Sutorectus tentaculatus	Cobbler wobbegong, cobbler carpet s	14. Bottom, inshore	200-300	Brown, dark saddles and spots	Suto-rectus = seamed muscle, possibly named for the rows of 'stitching' in form of rows of warty lumps down back.

Hemiscylliidae (Bamboosharks, long-tailed carpetsharks)

Scientific name	Common names	Distribution*	Size (cm)*	Colour	Notes
Chiloscyllium arabicum	Arabian carpet-shark	11. Bottom, inshore and offshore, shallow to moderately deep waters	70	Fawn	Chilo-scyllium = lipped dogfish. Members of this genus have flabby lower lips. So far only found in the Persian Gulf.
Chiloscyllium caerulopunctatum	Bluespotted bambooshark, bluespotted catshark	10	67 +	Dusky, pale blue spots	Rare, little known. Until seen recently in fish hauls in Madagascar, known only from one specimen (1914).
Chiloscyllium griseum	Grey bamboo-shark	11,12,13. Bottom, inshore	74	Fawn, dark/light bands on juveniles	Sluggish and unafraid; can be caught by hand.
Chiloscyllium indicum	Slender bamboo-shark	11,13,16. Bottom, inshore	65	Fawn, with dark dots and smudges	Spanish and French names include 'elegant' – appropriate for its svelte body and markings.
Chiloscyllium plagiosum	Whitespotted bambooshark	11,12,13. Bottom, inshore	95	Chocolate, dark bands, white spots	Like several of the family, this shark does very well in captivity.
Chiloscyllium punctatum	Brownbanded bambooshark, brown-spotted cat s	11,12,13,14. Bottom, inshore	104	Fawn/russet, dark spots on juveniles	Can live up to half a day out of water. Common on reefs and in ports.
Hemiscyllium freycineti	Indonesian speckled carpetshark, Freycinet's s	13. Bottom, shallow	46	Cream to tan, rusty spots, cream below	The family all have sturdy paired fins that help them to clamber on reefs.
Hemiscyllium hallstromi	Papuan epaulette shark	13. Bottom, inshore	75 +	Tan, large and small dark spots	Little known.
Hemiscyllium ocellatum	Epaulette shark	13,14. Bottom, shallow tropical waters	107	Brown, dark spots, white-edged fins	Abundant on Great Barrier Reef. Harmless (like all this family). Will happily search for food at a reef-walker's feet.
Hemiscyllium strahani	Hooded carpet-shark	13. Bottom, inshore	75	Black 'hood' on head, white spots	Little known.
Hemiscyllium trispeculare	Speckled carpet-shark, speckled catshark	14. Shallow tropical waters	64	Tan, brown spots, bands on tail	Common on coral reefs.

Stegostomatidae (Zebra sharks)

Scientific name	Common names	Distribution*	Size (cm)*	Colour	Notes
Stegostoma fasciatum	Zebra shark, monkey-mouthed s, leopard s	10,11,12,13,14,16. Bottom, shallow, tropical	354 ♀201	Tawny, with dense dark spots	Has prominent ridges on body. Tail is half the total length. Docile. An attractive and popular aquarium species. Lays eggs in horny cases with tufts of 'hair' to anchor them.

Brownbanded bambooshark
Chiloscyllium punctatum
104 cm

Zebra shark
Stegostoma fasciatum
354 cm

Nurse shark
Ginglymostoma cirratum
304 cm

*For note on size, and distribution map, see box p154.

Scientific name	Common names	Distribution*	Size (cm)*	Colour	Notes
Ginglymostomatidae (Nurse sharks)					
Ginglymostoma brevicaudatum	Short-tail nurse shark, nurse s	10. Bottom, inshore	75	Dark brown, paler below	Ginglymo-stoma – hinged mouth. All the family suck prey into their large mouth cavities, and all have remarkably strong skin that makes excellent leather.
Ginglymostoma cirratum	Nurse shark	2,3,4,5,6,9. Bottom, shallow inshore	304 ♀247 ♂241	Mustard/grey, often with dark spots	Large, clumsy, sluggish; generally inoffensive but will bite fast if provoked, and jaws clamped on victims have had to be pried aprart. Is not to be confused with the grey nurse shark of Australia (see sand tiger shark, this page). Nocturnal. Sometimes rests during day in groups of as many as 36, even piled on each other. A popular and hardy aquarium species; some have been kept for 25 years. Is taken for food, and for its skin, and in Brazil fishermen use its earstones (otoliths) as a diuretic.
Nebrius ferrugineus	Tawny nurse shark, Madame X, spitting s, giant sleepy s	10,11,13,14,16. Bottom, inshore, very shallow to moderately deep tropical waters	320 ♀260 ♂250	Tawny	Sucks prey into its mouth, and can reverse this action to spit when caught. Grunts between spittings. Nocturnal; rests in piles in crevices and caves. Thrives in captivity; will bite if provoked.
Rhiniodontidae (Whale sharks)					
Rhiniodon typus	Whale shark	2,3,4,5,6,9,10,11,12,13,14,16. Tropical and warm temperate waters, oceanic and inshore	1370 ♀800 ♂900	Charcoal, yellow spots, white below	The largest shark and the largest fish in the world (larger than small whales). Rhini-odon means file tooth; it has over 300 bands of minute teeth. Is a filter feeder, eating tiny plankton/small fishes, but will also take garbage: the stomach of one revealed a boot, a tin bucket, a wallet and part of an oar. Sometimes groups in hundreds. So inoffensive that people have walked on their backs.
LAMNIFORMES (Mackerel sharks) **Odontaspididae (Sandtiger sharks)**					
Eugomphodus taurus	Sandtiger shark, grey nurse s, sand s, ragged-tooth s, spotted ragged-tooth s	4,5,6,8,9,10,12,13,14. All water levels, surf zone and reefs	318 ♀260 ♂238	Mouse-grey, yellow spots, white below	Has a bad reputation as man-eater in Australia but this may be due to confusion with other sharks as it is inoffensive unless provoked. Popular aquarium species because it looks fierce. Inside the mother, several embryos develop at different stages; the largest, with well developed teeth, eats eggs and smaller embryos; one embryo per uterus survives, to be born about 100 cm (39 in) long. Can swallow air and hold it in stomach for neutral buoyancy. Groups sometimes herd schools of prey. IGFA game fish.
Eugomphodus tricuspidatus	Indian sandtiger	11. Inshore and offshore waters	370 +	Mouse-grey, paler below	This fish is possibly the sandtiger shark described above. The type specimen taken in 1878 was lost.
Odontaspis ferox	Smalltooth sand-tiger, shovel-nose nose shark	1,2,7,8,9,10,12,14,15,16. Bottom, moderate to deep waters	360 ♂275	Charcoal, russet spots, paler below	Despite the name *ferox*, this shark has never been accused of attacking people.
Odontaspis noronhai	Bigeye sandtiger	4,9. Deep	360	Dark brown	Until recently known only from one specimen (1955). Has been taken from between 600 and 1000 m (2000 and 3300 ft) down.
Mitsukurinidae (Goblin sharks)					
Mitsukurina owstoni	Goblin shark, elfins	4,7,9,10,12,14. Bottom, deep offshore, and shallow inshore waters	335 ♂293	Pink-white	Until found off Japan in 1898, believed extinct for 100 million years. Has weird snout overhanging protruding jaws.

Goblin shark
Mitsukurina owstoni
335 cm

Sandtiger shark
Eugomphodus taurus
318 cm

Whale shark
Rhiniodon typus
1370 cm

Scientific name	Common names	Distribution*	Size (cm)*	Colour	Notes
Pseudocarchariidae (Crocodile sharks)					
Pseudocarcharias kamoharai	Crocodile shark	2,9,10,12,13,16. Oceanic, pelagic; sometimes bottom inshore	110 ♀96 ♂92	Grey, paler below, white fin edges	Is a cannibal inside the mother like the sandtiger shark but, mysteriously, two embryos per uterus survive. Can bite very hard when captured. Oily liver gives neutral buoyancy.
Megachasmidae (Megamouth sharks)					
Megachasma pelagios	Megamouth shark	16. Deep, oceanic waters	446	Dusky brown to black	Known only from two specimens. Finding a new giant species that was also a new genus and new family was the most exciting shark discovery this century (off Hawaii, 1976). It has an enormous luminous mouth with 100 + rows of teeth. Lives midwater.
Alopiidae (Thresher sharks)					
Alopias pelagicus	Pelagic thresher, smalltooth thresher	2,10,11,12,13,14,16. Oceanic, sometimes inshore, surface to moderate depths	330 ♀297 ♂276	Grey, white below	All thresher sharks have tails nearly as long as the rest of the body, and they are cannibals within the uterus. All IGFA fishes.
Alopias superciliosus	Bigeye thresher	2,4,5,6,7,8,9,10,11,12,14,15,16. Inshore, and surface to deep offshore waters	461 ♀392 ♂335	Charcoal/ purple, large green eyes	Threshers hunt cooperatively, encircling schools of prey and shepherding them with their long tails, used to stun and kill.
Alopias vulpinus	Thresher shark, whiptail s, fox s, thintail thresher	Worldwide in temperate to tropical seas, pelagic, surface to deep waters	549 ♀462 ♂359	Sooty grey, offwhite below	Sometimes leaps out of water. Threshers have been seen to use their tails to 'down' birds skimming the surface.
Cetorhinidae (Basking sharks)					
Cetorhinus maximus	Basking shark, bone s, elephant s, sailfish s	1,2,3,4,6,7,8,9,12,13,14,15. Surface, inshore and offshore	1500 ♀890 ♂650	Deep blue to charcoal, pale below	The second largest shark; can weigh about 4 tonnes. A filter-feeder, it may filter more than 2000 tonnes of water an hour. Often seen basking with dorsal fin above water, even upside down. Pairs or threes swimming nose-to-tail were probably basis for sea monster legends. Gestation may be 3 years 6 months. Is inoffensive unless provoked. One liver can yield 750 litres (165 gallons) of oil, the record being twice as much; the oil once used in tanning and in lamps is now taken for squalene.
Lamnidae (Mackerel sharks, porbeagles, white sharks)					
Carcharodon carcharias	Great white shark, uptail, Tommy, death shark, maneater, blue pointer, white pointer, white death	All tropical, sub-tropical and warm temperate seas, in surf and shallow bays	800 ♂345	Slaty brown or charcoal, white below	The most feared and fearsome of all sharks. Attacks boats, sometimes persisting until boat sinks. Can swim in bursts of high speed. Sometimes leaps out of water, has been known to leap into a boat. Serrated, razor-sharp teeth and powerful jaws allow it to eat almost any animal, including turtles, seals. Will eat carrion like dead baleen whales, garbage. Jaws and teeth fetch high prices as curios. IGFA game fish.
Isurus oxyrinchus	Shortfin mako, bonito s, blue pointer	All tropical and warm temperate seas, surface to moderately deep	394 ♀337 ♂240	Metallic blue, snow white below	The fastest shark and one of the fastest of fishes. Possibly the most prized game fish; puts up furious resistance. Leaps spectacularly, even into boats. Few reliable reports of attacks on humans but when stimulated can be savage and attack swiftly with huge grasping teeth. Considered dangerous. Like other family members, a cannibal within mother. Excellent eating. Jaws and teeth are made into ornaments.

Crocodile shark
Pseudocarcharias kamoharai
110 cm

Megamouth shark
Megachasma pelagios
446 cm

Pelagic thresher shark
Alopias pelagicus
330 cm

*For note on size, and distribution map, see box p154.

Scientific name	Common names	Distribution*	Size (cm)*	Colour	Notes
Isurus paucus	Longfin mako	6,9,10,16. Deep, oceanic, tropical and warm temperate seas	417 ♀331	Blue-black, white below	Mako is a Maori word. Not recorded as having attacked people or boats but its large teeth make it dangerous. Makos can maintain body temperature higher than surrounding water.
Lamna ditropis	Salmon shark, Pacific porbeagle	1,2,12 and Bering Sea, surface to moderate depths, inshore and oceanic	305 ♂210	Dusky, pale below with dark marks	Di-tropis = two-keeled. All mackerel sharks have broad keeled tails but the two Lamna sharks have small extra keel. Salmon are a favoured food of this one.
Lamna nasus	Porbeagle, mackerel shark	3,4,6,7,8,9,10,14,15, also subantarctic waters and Barents Sea, all levels	300 ♀185 ♂240	Slaty blue/ charcoal, white below	Often found in colder water than suits other sharks. A strong swimmer, swift and voracious. Like makos, can keep its body temperature several degrees higher than water around. IGFA game fish.

CARCHARHINIFORMES (Ground sharks)
Scyliorhinidae (Catsharks)

Scientific name	Common names	Distribution*	Size (cm)*	Colour	Notes
Apristurus atlanticus	Atlantic ghost catshark	9. Bottom, deep	25	Brown, dark edges to fins and gills	The catshark genus is one of the largest among sharks.
Apristurus brunneus	Brown catshark	1,2. Bottom, deep	68 ♀46 ♂53	Chocolate brown, pale fin edges	Little is known of any of the catsharks.
Apristurus canutus	Hoary catshark	5. Bottom, deep	45 ♀43 ♂43	Charcoal, darker edges to fins	All catsharks have strikingly flabby bodies and thin skins.
Apristurus herklotsi	Longfin catshark	13. Bottom, deep	31 +	Brown	Known from only one specimen (Philippines, 1934).
Apristurus indicus	Smallbelly catshark	10,11. Bottom, deep	34 +	Mouse-grey	
Apristurus investigatoris	Broadnose catshark	11. Bottom, deep	26 +	Brown	
Apristurus japonicus	Japanese catshark	12. Bottom, deep	71 ♂68	Dusky	Abundant.
Apristurus kampae	Longnose catshark	1,2. Bottom, deep	52	Dusky/black, white fin edges	Like several of the catsharks, found in very deep water (1900 m – 6200 ft down).
Apristurus laurussoni	Iceland catshark	5,6,7,9. Bottom, deep	68 ♀67 ♂68	Dusky to black	None of this family have any colour pattern.
Apristurus longicephalus	Longhead catshark	12. Bottom, deep	37 +	Charcoal	Has a very long snout. Known from only one specimen (1975).
Apristurus macrorhynchus	Flathead catshark	12. Bottom, deep	66	Dun, paler below and pale fins	Most members of this family lay eggs in sturdy egg-cases anchored with tendrils.
Apristurus maderensis	Madeira catshark	9. Bottom, deep	68		Several new members of this genus have been discovered in the last 15 years. It may turn out to be largest shark genus.
Apristurus manis	Ghost catshark	6,7. Bottom, deep	85	Ash	Odd in profile: body very thick in the middle tapering to a narrow snout.
Apristurus microps	Smalleye catshark	9. Bottom, deep	54 +	Dusky or purplish black	Has a fuzzy texture to its skin, like several members of this genus.

Basking shark
Cetorhinus maximus
1500 cm

Great white shark
Carcharodon carcharias
800 cm

Porbeagle shark
Lamna nasus
300 cm

Scientific name	Common names	Distribution*	Size (cm)*	Colour	Notes
Apristurus nasutus	Largenose catshark	3. Bottom, deep	59 ♂55	Brown to charcoal	
Apristurus parvipinnis	Smallfin catshark	4,5. Bottom, deep	52	Black	Common in Gulf of Mexico.
Apristurus platyrhynchus	Spatulasnout catshark	12. Deep	80	Dark brown	
Apristurus profundorum	Deepwater catshark	6. Deep	51 +	Dusky or black	Has been found 1600 m (5250 ft) down. Since it was named, others in this genus have been found living in deeper waters.
Apristurus riveri	Broadgill catshark	5. Bottom, deep	46 ♀40 ♂44	Chocolate brown	Males have much larger teeth, jaws and mouths than the females, possibly so the female can be grasped during mating.
Apristurus saldanha	Saldanha catshark	9. Deep	81	Blue-grey	
Apristurus sibogae	Pale catshark	13. Deep	21 +	Reddish white	Known from only one specimen (Makassar Straits, 1913).
Apristurus sinensis	South China catshark	13. Bottom, deep	50 +	Dark brown	Known from only one immature specimen (South China Sea, 1981).
Apristurus spongiceps	Spongehead catshark	13,16. Bottom, deep	50	Dark brown	Unique in having pleats and grooves around gills and throat. Known from only two specimens.
Apristurus stenseni	Panama ghost catshark	2. Deep	23 +	Charcoal	Has an enormous mouth.
Apristurus verweyi	Borneo catshark	13. Deep	30 +	Chocolate	Known from only one specimen (1934).
Asymbolus analis	Australian spotted catshark	14. Bottom, inshore, and at moderate depths offshore	61 ♀57 ♂55	Fawn, russet spots, pale below	Lives in temperate waters off southern Australia.
Asymbolus vincenti	Gulf catshark	14. Bottom, moderate to deep waters	61 ♀49 ♂51	Dusky, dense white spots, white below	Found in southern waters of Australia.
Atelomycterus macleayi	Australian marbled catshark	14. Bottom, shallow	60 ♀51 ♂48	Fawn with grey saddles, black spots	Sometimes found in water merely 0.5 m (20 in) deep.
Atelomycterus marmoratus	Coral catshark	11,12,13. Shallow	70 ♀53 ♂54	Fawn, dark streaks and white spots	Lives in crevices in coral reefs. Does well in captivity.
Aulohalaelurus labiosus	Blackspotted catshark	14. Shallow	67 ♀67 ♂58	Fawn, grey saddles, black dots	Found off Western Australia only.
Cephaloscyllium fasciatum	Reticulated swellshark	13,14. Bottom, deep	42 +	Dun, dark brown saddle, dark spots	When disturbed, swellsharks can inflate their stomachs with water or air to become almost spherical.
Cephaloscyllium isabellum	Draughtsboard shark, Isabel's swellshark	12,15. Bottom, moderate depths	100 + ♀86 ♂69	Fawn, dark saddles with dark blotches	Colouring in the New Zealand form is more distinctly chessboard-like than in Taiwanese and Japanese forms.
Cephaloscyllium laticeps	Australian swell-shark, draught-board s	14. Bottom, shallow inshore to moderately deep waters	97 +	Grey-russet, dark patches, pale below	Called 'Sleepy Joe' because it, like all swellsharks, is sluggish.

Smallfin catshark
Apristurus parvipinnis
52 cm

Spongehead catshark
Apristurus spongiceps
50 cm

Australian spotted catshark
Asymbolus analis
61 cm

*For note on size, and distribution map, see box p154.

Scientific name	Common names	Distribution*	Size (cm)*	Colour	Notes
Cephaloscyllium nascione	Whitefinned swellshark	14. Bottom, moderately deep waters	100 ♂80+	Fawn, dark saddles, fins white-edged	All swellsharks except the Indian have jaws jointed so teeth are visible when viewed from below.
Cephaloscyllium silasi	Indian swellshark	11. Bottom, deep	36	Fawn with seven dark saddles	A dwarf swellshark.
Cephaloscyllium sufflans	Balloon shark, swellshark	10. Bottom, moderate and deep waters	106 ♀96	Pale with seven darker saddles	
Cephaloscyllium ventriosum	Swellshark	1,2,3. Bottom, shallow and deep waters	100 ♂83	Tan, dark and light spots and patches	Nocturnal. Groups sometimes rest heaped on top of each other. Thrives and has laid eggs in captivity. Young emerging from egg cases have two rows of large denticles used like ratchets to help them out.
Cephalurus cephalus	Lollipop catshark, head shark	2. Bottom, deep	28 ♀21 ♂21	Light/dark brown, green eyes	Scientific name means head-headed; very large gill region is probably an adaptation to life at bottom with low oxygen.
Galeus arae	Roughtail catshark, marbled catshark	4,5,6. Bottom, deep	43 ♀34 ♂31	Tawny, white outlined saddles	It is possible that the Caribbean subspecies lays eggs while the continental slopes subspecies bears live young.
Galeus boardmani	Australian sawtail catshark	14. Bottom, moderate to deep waters	61 ♂54	Pale grey, white outlined dark bands	Sawtail because of the rows of denticles on the edges of the tail.
Galeus eastmani	Gecko catshark	12,13. Bottom, deep	50 ♀38 ♂33	Dark patches, white tipped fins	'Gecko' because of the slender body. Egg-cases measure about 1.6 × 6 cm (0.6 × 2.4 in).
Galeus melastomus	Blackmouth catshark, blackmouthed dogfish	7,8,9. Bottom, moderate to deep waters	90 ♀66 ♂54	Tawny, dark patches, off-white below	Galeus = dogfish (Greek). The French common name means Spanish dog.
Galeus murinus	Mouse catshark	7. Bottom, deep	63	Brown, paler below	Found off Iceland and the Faroes Islands only.
Galeus nipponensis	Broadfin sawtail catshark	12. Bottom, deep	65 ♀54 ♂54	Light mouse with dark saddles	The males' anal fin is much shorter than the females', a development probably related to the very long claspers.
Galeus piperatus	Peppered catshark	2. Bottom, deep	30 ♀28 ♂28	Tan, black dots, ash below	Found only in the northern Gulf of California.
Galeus polli	African sawtail catshark	9. Bottom, deep	42 ♀41 ♂37	White-edged dark blotches on back	
Galeus sauteri	Blacktip sawtail catshark	12. Bottom, at moderate depths offshore	45 ♀43 ♂37	Brown, black tips to fins and tail	
Galeus schultzi	Dwarf sawtail catshark	13. Bottom, deep	30 ♀28 ♂25	Saddles on dorsal fins, bands on tail	One of the smallest of sharks.
Halaelurus alcocki	Arabian catshark	11. Bottom, deep	30	Charcoal/ash, white tips to fins	Known from only one specimen (1913).

Australian swellshark
Cephaloscyllium laticeps
97 cm

Swellshark
Cephaloscyllium ventriosum
100 cm

Australian sawtail catshark
Galeus boardmani
61 cm

Scientific name	Common names	Distribution*	Size (cm)*	Colour	Notes
Halaelurus boesemani	Speckled catshark	10,11,13,14. Bottom, moderate depths	48 ♀45 ♂45	Brown, black speckles, grey patches	
Halaelurus buergeri	Blackspotted catshark	12,13. Bottom, moderate depths	49 ♀45 ♂39	Fawn, black spots, dark back patches	Like several relatives, the female retains several egg-cases inside uterus until embryos are well developed before laying the eggs.
Halaelurus canescens	Dusky catshark	3. Bottom, deep	70 ♀62 ♂62	Dusky	
Halaelurus dawsoni	New Zealand catshark	15. Bottom, deep	45 ♀40 ♂40	Fawn/grey, pale below, white spots	Found only off Auckland Island and the southern tip of New Zealand.
Halaelurus hispidus	Bristly catshark	11. Bottom, deep	29 ♀25 ♂25	Fawn/off-white, often stripes/spots	A very small catshark.
Halaelurus immaculatus	Spotless catshark	13. Bottom, deep	76 ♀75 ♂71	Ochre	Recently discovered (described in 1982).
Halaelurus lineatus	Lined catshark, banded catshark	10. Inshore, surf zone to deep waters	56 ♀49 ♂52	Fawn, dense flecks, 26 dark saddles	Has an uptilted end to snout. Easily kept in captivity.
Halaelurus lutarius	Mud catshark, brown catshark	10. Bottom, deep	39 ♀35 ♂32	Dun, paler below	Lutarius = muddy; named because of its preference for muddy sea floors.
Halaelurus natalensis	Tiger catshark	10. Bottom, inshore to moderate depths	47 ♀47 ♂43	Tawny, ten dark saddles, cream below	Taken for sport.
Halaelurus quagga	Quagga catshark	10,11. Bottom, moderately deep tropical offshore waters	35 ♂31	Pale, zebra-like stripes, paler below	The quagga, now extinct, was a member of the horse family, striped like a zebra.
Haploblepharus edwardsii	Puffadder shyshark	10. Bottom, surf zone to moderately deep offshore waters	60 ♀50 ♂53	Cream, white speckles, grey saddles	'Shyshark' because when caught they hide their eyes with their tails.
Haploblepharus fuscus	Brown shyshark	10. Shallow, inshore	73 ♀66 ♂66	Reddish brown, vague saddles, white below	Does well in captivity.
Haploblepharus pictus	Dark shyshark	9,10. Bottom, shallow	56 ♀53 ♂56	Mustard brown, dark saddles, paler below	All the shysharks are sport fishes and prey to anglers.
Holohalaelurus punctatus	African spotted catshark	10. Bottom, moderately deep waters	34 ♀25 ♂31	Mustard, brown dots, white fin dot	Unlike most sharks, males of this genus are larger than females.
Holohalaelurus regani	Izak catshark, Izak, shy eye s	9,10. Bottom, moderate to deep waters	61 ♀49 ♂56	Dark brown, dark dots and light lines	Another species that hides its eyes with its tail when caught.
Parmaturus campechiensis	Campeche catshark	5. Bottom, deep		Brown	Known only from one immature specimen (Bay of Campeche, Gulf of Mexico, 1979).
Parmaturus melanobranchius	Blackgill catshark	13. Bottom, deep	85	Pale brown	All members of this genus have flabby bodies.

New Zealand catshark
Halaelurus dawsoni
45 cm

Tiger catshark
Halaelurus natalensis
47 cm

Brown shyshark
Haploblepharus fuscus
73 cm

*For note on size, and distribution map, see box p154.

Scientific name	Common names	Distribution*	Size (cm)*	Colour	Notes
Parmaturus pilosus	Salamander shark	12. Bottom, deep	64 ♀61 ♂56	Brown	Salamanders are tailed amphibians with long bodies. Pilosus – hairy, perhaps for the crest of denticles on tail fin.
Parmaturus xaniurus	Filetail catshark	1,2. Bottom, deep	55 ♀51 ♂41	Dusky brown to black	Very large gill region, probably an adaptation to living deep down where oxygen levels are very low.
Pentanchus profundicolus	Onefin catshark	13. Bottom, deep	50	Colour not known	Has only one dorsal fin, like the frilled and cow sharks. Known from only one specimen (Philippines, 1913).
Poroderma africanum	Striped catshark	9,10. Bottom, surf zone to moderate depths inshore	101 ♀80 ♂84	Ash, dark longitudinal stripes	Often found in caves and among rocks. Does well in aquariums. A game fish.
Poroderma marleyi	Barbeled catshark, black-spotted catshark	10. Moderately deep waters	65 ♂58	Ash, large dark spots on back	Rare.
Poroderma pantherinum	Leopard catshark	9,10. Bottom, from surf zone to moderate depths	84 ♀66 ♂70	Spots/rings in longitudinal rows	Does well in captivity. A game fish.
Shroederichthys bivius	Narrowmouthed catshark	3,4. Moderate depths, inshore and offshore waters	70 ♀40 ♂53	Mouse-grey, dark saddles some spots	Males' teeth are twice as long as females', and their mouths are much longer than females'.
Schroederichthys chilensis	Redspotted catshark	3. Bottom, close inshore	62 ♂59	Pale brown, dark saddles and spots	
Schroederichthys maculatus	Narrowtail catshark	5. Bottom, deep tropical waters	34 ♀34 ♂30	Light brown/ grey, yellow-white spots	Strikingly slender-bodied.
Schroederichthys tenuis	Slender catshark	4. Bottom, deep tropical waters	70	Fawn, darker saddles edged with spots	Occurs off the Amazon River mouth.
Scyliorhinus besnardi	Polkadot catshark	4. Bottom, moderately deep waters	47	Pale, sparse black spots	Scylio-rhinus = dogfish-nose.
Scyliorhinus boa	Boa catshark	5. Bottom, deep	54	Fawn/grey, saddles, rows of black dots	'Boa' because of its colouring.
Scyliorhinus canicula	Small-spotted catshark, sandy dogfish, rough hound	7,8,9. Bottom, shallow to moderately deep waters	100	Sandy, dense dark spots, cream below	Abundant. Eaten as 'huss' and 'flake' in Britain. Thrives in captivity.
Scyliorhinus capensis	Yellowspotted catshark	10. Bottom, inshore and offshore, moderate to deep waters	122 ♀77 ♂83	Grey, dark saddles, yellow spots	Like several shark species, believed to live deeper in warmer waters.
Scyliorhinus cervigoni	West African catshark	9. Bottom, moderate to deep tropical waters	76	Vague dark saddles/spots on flanks	Believed to lay eggs in cases 7 × 3 cm (3 × 2 in).
Scyliorhinus garmani	Brownspotted catshark	13.	24 or 36 +	Large brown spots, vague saddles	Little known – even size of adults is uncertain.

Izak catshark
Holohalaelurus regani
61 cm

Barbeled catshark
Poroderma marleyi
65 cm

Narrowtail catshark
Schroederichthys maculatus
34 cm

Scientific name	Common names	Distribution*	Size (cm)*	Colour	Notes
Scyliorhinus haeckeli	Freckled catshark	4. Bottom, moderate to deep tropical waters	35+	Tiny black spots on back, dark saddles	Little known.
Scyliorhinus hesperius	Whitesaddled catshark	5. Bottom, deep	47+	Large white spots and dark saddles	Hesperius = western; named for the region of the Atlantic Ocean in which it is found.
Scyliorhinus meadi	Blotched catshark	5,6. Bottom, deep	49+	Ash with 7 or 8 dusky saddles	Rare.
Scyliorhinus retifer	Chain catshark, chain dogfish	5,6. Bottom, deep	47 ♀41 ♂39	Yellow-green eyes, cream, brown marks	Stomachs have revealed small pebbles, possibly taken as ballast.
Scyliorhinus stellaris	Nursehound, large-spotted dog-fish, bull huss	7,8,9. Bottom, very shallow inshore to moderately deep offshore	162 ♀125 ♂125	Mouse-brown, dark and white spots	Another shark sold as 'flake' in British fish shops. Skin is sometimes made into leather called rubskin.
Scyliorhinus torazame	Cloudy catshark	12,13. Shallow, inshore to moderately deep waters	48 ♀39+ ♂44	Dusky, dark saddles, some light spots	Females deposit eggs in a nursery area.
Scyliorhinus torrei	Dwarf catshark	5,6. Bottom, deep	32 ♀26 ♂25	Fawn, dark blotches, white spots	Occurs in a limited area, off Florida, the Bahamas and northern Cuba.

Proscyllidae (Finback catsharks)

Scientific name	Common names	Distribution*	Size (cm)*	Colour	Notes
Ctenacis fehlmanni	Harlequin catshark	10. Bottom, deep tropical waters	46	Russet bars, blotches and spots	Known from only one specimen (Somalia, 1973).
Eridacnis barbouri	Cuban ribbontail catshark	5. Bottom, deep subtropical waters	34 ♀28 ♂27	Dun or ash, off-white below	Very small and slender.
Eridacnis radcliffei	Pygmy ribbontail shark	10,11,13. Bottom, deep tropical waters	24 ♀19 ♂20	Brown, dark marks on fins and tail	The second-smallest shark. Produces one or two young, about 11 cm (4 in) at birth – remarkable as females mature at about 16 cm (6 in).
Eridacnis sinuans	African ribbontail catshark	10. Bottom, deep dark	37 ♂29	Mouse-grey, bands on dorsal fins	
Gollum attenuatus	Slender smooth-hound	15. Bottom, deep	101 ♀101 ♂94	Mouse-grey, paler below	Little known and uncommon.
Proscyllium habereri	Graceful catshark	12,13. Bottom, moderately deep waters	65 ♀51 ♂42	Large and small dark spots	

Pseudotriakidae (False catsharks)

Scientific name	Common names	Distribution*	Size (cm)*	Colour	Notes
Pseudotriakis microdon	False catshark, dumb shark	6,8,9,10,13,16. Bottom, deep	295 ♀253 ♂234	Dusky	Micr-odon = tiny-toothed, but there are about 200 rows of teeth in each jaw, and the mouth is very large. Sluggish.

Leptochariidae (Barbelled houndsharks)

Scientific name	Common names	Distribution*	Size (cm)*	Colour	Notes
Leptocharias smithii	Barbeled houndshark	9. Bottom, tropical inshore waters	82 ♀70 ♂67	Ash or mouse-grey	Abundant off river mouths. Males' front teeth much larger than females', perhaps related to mating behaviour. Found in water with temperature between 20° and 27°C (68° and 81° F).

Nursehound
Scyliorhinus stellaris
162 cm

Harlequin catshark
Ctenacis fehlmanni
46 cm

False catshark
Pseudotriakis microdon
295 cm

*For note on size, and distribution map, see box p154.

Scientific name	Common names	Distribution*	Size (cm)*	Colour	Notes
Triakidae (Houndsharks, smooth-hounds, topes, whiskery sharks)					
Furgaleus macki	Whiskery shark	14. Bottom, moderately deep waters	160 ♀121 ♂122	Grey, paler below	'Whiskery' because of moustache-like nasal barbels.
Galeorhinus galeus	Tope shark, school s, soupfin s, vitamin s	1,3,4,7,8,9,10,14,15. Bottom, surf zone to far offshore	195 ♀162 ♂145	Pale bluish grey to purple-black	Galeo-rhinus galeus means shark shark shark. An important food fish. Not dangerous but will bite when caught. A game fish. Tagging shows it can live over 40 years. Can travel 56 km (35 miles) in one day. Formerly so much used for liver oil it was called the oil shark. Can produce 50 pups in one litter.
Gogolia filewoodi	Sailback houndshark	13. Bottom, moderately deep	74	Mouse-grey	Known from only one specimen (Papua New Guinea, 1973).
Hemitriakis japanica	Japanese topeshark	12,13. Close inshore and moderately deep offshore waters	120 ♀106 ♂92	White edges to fins	
Hemitriakis leucoperiptera	Whitefin topeshark	13. Moderately deep inshore waters	96	White edges to fins	Leuco-periptera = white-winged.
Hypogaleus hyugaensis	Blacktip tope, lesser soupfin shark	10,11,12. Bottom, tropical and subtropical moderately deep waters	127 ♀118 ♂119	Ash, off-white below	Like almost every shark occurring near Japan, this fish is an important food source.
Iago garricki	Longnose hound-shark	16. Bottom, deep tropical waters	75 ♀63	Colour not known	So far, found only off Vanuatu. At birth, the young are 23 cm (11 in) long.
Iago omanensis	Bigeye hound-shark	11. Bottom at considerable depths, tropical waters	58 ♀49 ♂33	Mouse-grey, often black fin edges	Large gill region is an adaptation to life in water with little oxygen and a lot of salt. Males are two-thirds the size of females, and weigh one-sixth as much.
Mustelus antarcticus	Gummy shark, Sweet William, flake, smooth dog s	14. Bottom, inshore to moderate depths offshore	157 ♀80 ♂68	Grey, many small white spots	Ant-arcticus = anti-northern, i.e. southern. 'Gummy' refers to its flattened teeth. Has a strong odour.
Mustelus asterias	Starry smooth-hound, stellate smooth- hound	7,8,9. Bottom, shallow inshore to moderate depths offshore	140 ♀85 ♂81	Ash/charcoal, white spots, paler below	Called 'Stinkard' by Irish fishermen.
Mustelus californicus	Grey smooth-hound	1,2. Bottom, shallow inshore and offshore waters	124 ♀97 ♂88	Dark brown/grey, paler below	Will enter muddy bays. Does well in captivity.
Mustelus canis	Dusky smooth-hound, smooth dogfish	4,5,6. Bottom, shallow inshore to deep offshore waters	150 ♀106	Mouse, olive-grey/brown, white below	Can change colour to blend with background. Abundant off the United States, second only to the piked dogfish. Enters lower reaches of rivers but probably cannot live long in fresh water. Often used in research as it thrives in captivity. A game fish.
Mustelus dorsalis	Sharp-tooth smooth-hound	2,3. Bottom, inshore, tropical waters	64 ♀53	Mouse-grey or grey, paler below	Mustelus = weasel, because of its long snout.
Mustelus fasciatus	Striped smooth-hound	4. Bottom, inshore	125	Mouse-grey, paler below	Juveniles have dark stripes.
Mustelus griseus	Spotless smooth-hound	12,13. Bottom, inshore	101 ♀90 ♂77	Mouse-grey or charcoal, paler below	Survives in captivity.

Barbeled houndshark
Leptocharias smithii
82 cm

Tope shark
Galeorhinus galeus
195 cm

Dusky smooth-hound
Mustelus canis
150 cm

Scientific name	Common names	Distribution*	Size (cm)*	Colour	Notes
Mustelus henlei	Brown smooth-hound	1,2,3. Bottom, shallow inshore to moderate depths offshore	95 ♀72 ♂63	Copper, white below	Abundant. Often found in shallow muddy bays and around oyster beds. Does well in captivity. Active and agile.
Mustelus higmani	Smalleye smooth-hound	4. Bottom, close inshore to moderate depths offshore	64 ♀53 ♂46	Mouse-grey or grey, paler below	Tolerates brackish water.
Mustelus lenticulatus	Spotted estuary smooth-hound, white-spotted gummy shark, rig	15. Bottom, inshore to deep offshore waters	137 ♀117 ♂99	Mouse-grey or grey, dense white spots	The only New Zealand smooth-hound. Important to New Zealand fishing industry. A sport fish.
Mustelus lunulatus	Sicklefin smooth-hound	1,2. Bottom, inshore to deep offshore waters	170 ♂103	Olive-brown or grey, paler below	Abundant.
Mustelus manazo	Starspotted smooth-hound	10,12,13. Bottom, close inshore	117 ♀91 ♂81	Mouse-grey, white spots, paler below	Abundant. An important food fish in Japan. Survives in captivity.
Mustelus mento	Speckled smooth-hound	3,4. Shallow inshore to moderate depths offshore	130 ♀88 ♂70	Mouse-grey, white spots, paler below	
Mustelus mosis	Arabian smooth-hound, Moses smooth-hound, hardnose smooth-hound	10,11. Bottom, inshore and offshore waters	150 ♂85	Mouse-grey or grey, paler below	Does well in captivity.
Mustelus mustelus	Smooth-hound	7,8,9. Bottom, shallow inshore to deep offshore waters	164 ♀122 ♂101	Grey, often black spots, white below	Litters of 4 to 15 are about 39 cm (15 in) at birth. Abundant. An important food fish. Also taken by anglers.
Mustelus norrisi	Narrowfin smooth-hound, Florida dogfish	4,5. Bottom, close inshore to moderate depths	100	Grey, paler below	
Mustelus palumbes	Whitespotted smooth-hound	9,10. Bottom, inshore to deep offshore waters	120 ♀90 ♂82	Mouse-grey, white spots, paler below	Sometimes processed into 'biltong,' a dried meat. A sport fish.
Mustelus punctulatus	Blackspotted smooth-hound	8,9. Bottom, inshore black	85	Mouse-grey, spots, paler below	Its dorsal fins are fringed.
Mustelus schmitti	Narrownose smooth-hound	4. Bottom, moderately deep waters	74	Grey, often white spots, paler below	An important food fish in Argentina and Uruguay. Survives in captivity.
Mustelus whitneyi	Humpback smooth-hound	3. Bottom, moderately deep offshore waters	87	Mouse-grey or grey, paler below	
Scylliogaleus quecketti	Flapnose houndshark, flapnosed smooth-hound	10. Surf zone to close offshore	102 ♀91 ♂79	Grey, cream below	Has flaps to its nose that extend to the mouth. A sport fish.
Triakis acutipinna	Sharpfin houndshark	3.	102	Colour not known	Known only from two specimens taken off Ecuador.

Brown smooth-hound
Mustelus henlei
95 cm

Leopard shark
Triakis semifasciata
180 cm

Snaggletooth shark
Hemipristis elongatus
240 cm

*For note on size, and distribution map, see box p154.

Scientific name	Common names	Distribution*	Size (cm)*	Colour	Notes
Triakis maculata	Spotted houndshark	3. Inshore waters	180	Generally has many black spots	Tri-akis = three-pointed (refers to the teeth).
Triakis megalopterus	Sharptooth houndshark, gully s, Sweet William	9,10. Bottom, shallow waters, surf zone and in bays	174 ♀159	Charcoal, often many black spots	Named 'gully' for its liking for crevices. Does well in captivity. A sport fish. Used for biltong (a dried meat).
Triakis scyllium	Banded houndshark	12. Bottom, shallow	150	Indistinct black spots	Often found in bays and at river mouths, occasionally resting in groups. A successful aquarium species.
Triakis semifasciata	Leopard shark	1, 2. Bottom, shallow inshore	180 ♀149 ♂122	Golden brown, dark saddles, black spots	Abundant and harmless. One of the most hardy aquarium species. Often enters muddy bays with tide, leaving as it ebbs. Eats clam necks, possibly removing shells by vigorous shaking once the clam is pulled free.

Hemigaleidae (Weasel sharks)

Scientific name	Common names	Distribution*	Size (cm)*	Colour	Notes
Chaenogaleus macrostoma	Hooktooth shark	11,12,13. Shallow and moderately deep inshore waters	100 ♂82	Golden brown or ash	Hooked teeth and large mouth probably help it snatch and hold fish prey.
Hemigaleus microstoma	Sicklefin weasel shark, weasel s	11,12,13,14. Bottom, moderately deep tropical waters	97 ♀82 ♂77	Ash or copper, black/white fin tips	A specialist feeder on squid, cuttlefish and octopuses. Short mouth probably helps suction feeding.
Hemipristis elongatus	Snaggletooth shark	9,10,11,12,13,14. Shallow inshore to moderately deep offshore, tropical waters	240 ♀194 ♂132	Golden brown or grey	Potentially dangerous because of long teeth (those on lower jaw protrude when mouth is shut), its size and its occurrence in shallow water. Prized as food in India. Similar sharks, twice as large, lived 70 million years ago, worldwide.
Paragaleus pectoralis	Atlantic weasel shark	9. Shallow inshore to moderately deep offshore waters	138 ♀97 ♂97	Golden brown or grey, yellow bands	Also a specialist feeder on squid and octopuses. Small teeth probably suited to holding such soft and squirming prey.
Paragaleus tengi	Straight-tooth weasel shark	12,13. Inshore	88 ♂83	Ash	Little known.

Carcharhinidae (Requiem sharks)

Scientific name	Common names	Distribution*	Size (cm)*	Colour	Notes
Carcharhinus acronotus	Blacknose shark	4,5,6. Moderately deep inshore waters	200 ♀120	Mouse-grey, dark tips to fins/snout	In captivity, faced with intruders, arches its back, raises its head and lowers its tail, perhaps as a threat. Is prey to larger sharks. Harmless. IGFA game fish.
Carcharhinus albimarginatus	Silvertip shark, silvertip whaler	2,3,10,11,12,13,16. Surface to deep, inshore and offshore waters	300	Charcoal, white tips to fins	Plentiful around reefs and islands. Fast, large and bold. Could be dangerous.
Carcharhinus altimus	Bignose shark, Knopp's shark	2,3,4,5,6,8,9,10,11,16. Bottom, deep	300 ♀254 ♂241	Ash, darker fin tips, white below	Unlikely to be a danger since it lives deep down.
Carcharhinus amblyrhynchoides	Graceful shark	11,13,14. Inshore and out to sea, tropical waters	167+	Charcoal, white streak on flanks	Carcha-rhinus = jagged-shark, or jagged-file, referring to teeth or skin, perhaps both.
Carcharhinus amblyrhynchos	Grey reef shark, long-nosed black-tail shark	10,13,14,16. Bottom, reefs and lagoons, and at moderate depths offshore	255	Grey, tail black edge to tail fin	Curious but generally not dangerous. Has attacked people (at least one fatality); possibly spearfishers were mistaken for their catch. Another shark to adopt postures and swimming patterns to convey threat to intruders.

Straight-tooth weasel shark
Paragaleus tengi
88 cm

Graceful shark
Carcharhinus amblyrhynchoides
167 cm

Grey reef shark
Carcharhinus amblyrhynchos
255 cm

Scientific name	Common names	Distribution*	Size (cm)*	Colour	Notes
Carcharhinus amboinensis	Pigeye shark, Java shark	9,10,11,13,14. Shallow, surf zone, to moderate depths inshore	280 ♀210	Grey, darker fin tips, off-white below	To be respected because of size and large teeth, although no known attacks.
Carcharhinus borneensis	Borneo shark	13. Tropical inshore waters	70	Brown, off-white below	Rare and little known.
Carcharhinus brachyurus	Copper shark, New Zealand whaler, cocktail s	1,2,3,4,8,9,10,12,14,15. Surf zone to moderate depths offshore	292 ♀266 ♂233	Golden brown, cream streak, cream below	Dangerous. Has attacked people in South Africa, Australia and New Zealand. A sport fish.
Carcharhinus brevipinna	Spinner shark, smoothfanged s, longnose grey s	4,5,6,8,9,10,11,12,13,14. Shallow to moderately deep waters	278 ♀224 ♂196	Grey, black fin tips, paler below	Has small teeth but has attacked humans. Often swims fast upwards through schools of prey, snapping all around, emerging still spiralling in a leap out of water. IGFA game fish.
Carcharhinus cautus	Nervous shark	13,14. Shallow waters and reefs	150 ♀135	Mouse-grey, some black fin edges	Timid and probably not dangerous.
Carcharhinus dussumieri	Whitecheek shark, white-cheeked whaler shark	11,12,13. Inshore waters	100 ♀76 ♂73	Mouse-grey, off-white underside	Common, harmless.
Carcharhinus falciformis	Silky shark, sickle shark	2,3,4,5,6,9,10,11,12,13,15,16. Shallow inshore; all levels in open seas	330 ♀259 ♂243	Dusky to sooty, paler below	Called silky for its smooth skin. Is detested by tuna fishers for damage it wreaks on nets and catches; is named the net-eater shark in some parts of Pacific Ocean. Ranks with blue shark and oceanic whitetip shark as most common oceanic species. Not identified in any attacks but abundance and size make it a potential danger. Possibly the cause of deaths after ship sinkings or plane crashes at sea.
Carcharhinus fitzroyensis	Creek whaler	14. Inshore and offshore tropical waters	150	Grey, paler below	The whalers earned their name from their habit of despoiling whale catches in the early days of Australian and New Zealand whaling.
Carcharhinus galapagensis	Galapagos shark, grey reef whaler	2,3,5,9,10,16. Shallow inshore, surface to moderately deep near oceanic islands	370 ♀269 ♂231	Charcoal, white below	Aggressive and dangerous. When excited, will attract others to the area. Known to have killed at least one person.
Carcharhinus hemiodon	Pondicherry shark	11,13,14.	175	Grey, paler below	Little known. Possibly enters river estuaries but such reports are old and may not be reliable.
Carcharhinus isodon	Finetooth shark	4,5,6. Shallow inshore waters	189 ♀157 ♂149	Slate fading to white below	Is-odon means equal-toothed; has same numbers of teeth in upper and lower jaws.
Carcharhinus leucas	Bull shark, cub s, Ganges s, river s, Nicaragua s, Zambezi s, shovelnose, slipway grey s, square-nose s, Van Rooyen's s	2,3,4,5,6,9,10,11,12,13. Shallow inshore, rivers, bays and near wharves	340 ♀242 ♂228	Grey, off-white below	Can live for some time in fresh water; occurs 3700 km (2300 miles) from sea in upper Amazon. Once believed landlocked in Lake Nicaragua but now known to negotiate rapids and return to sea. Jaws and teeth allow it to take large prey, and it will eat almost anything. Eats many other sharks and rays including hammerheads, stingrays and even young of its own kind. Is abundant in many areas where large numbers of people use or enjoy sea or rivers. Is known to have caused several deaths but is not readily recognised so true number of fatalities may be much greater. Possibly the most dangerous tropical shark, even the most dangerous of all. Vastly dreaded in India where it attacks pilgrims in the holy river Ganges, and feeds on corpses consigned to sacred waters. Flesh, fins, hide, liver and carcases are used. IGFA game fish.

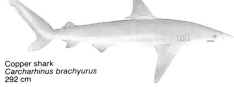

Copper shark
Carcharhinus brachyurus
292 cm

Silky shark
Carcharhinus falciformis
330 cm

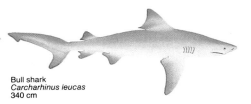

Bull shark
Carcharhinus leucas
340 cm

*For note on size, and distribution map, see box p154.

Scientific name	Common names	Distribution*	Size (cm)*	Colour	Notes
Carcharhinus limbatus	Blacktip shark, spotfin s, grey s, blackfin	1,2,3,4,5,6,8,9,10,11,12,13,14, 16. All tropical and subtropical waters inshore and out to sea	225 ♀184 ♂195	Charcoal, bronze tinged, white below	Often found in mangrove swamps and muddy estuaries. Leaps spinning from water after launching itself upwards through a school of prey. Becomes frenzied when competing for food, and fishermen have seen hundreds churning water to froth. Active, fast, sometimes aggressive. Few recorded attacks on people but is potentially dangerous.
Carcharhinus longimanus	Oceanic whitetip shark, white-tipped s	2,3,4,5,6,7,9,10,11,12,13,14, 16. All levels, tropical and subtropical waters	395 ♀225 ♂210	Mouse-grey, copper tinged, white below	Abundant and very dangerous. Hated by tuna fishers and formerly by whalers for harm it does to catches. Perseveres fearlessly when investigating prey. Slow-moving but can move in fast dashes.
Carcharhinus macloti	Hardnose shark, Maclot's shark	10,11,12,13. Inshore waters	100 ♀82 ♂75	Mouse-grey, pale fin edge white below	Was probably cause of many casualties when *Nova Scotia* was torpedoed off South Africa in World War II. Has a gristly beak inside the snout. Like many sharks, lives in sex-segrated populations.
Carcharhinus melanopterus	Blacktip reef shark, blacktip reef whaler, black shark	10,11,12,13,14,16. Shallow close inshore and close offshore	180 ♀113 ♂135	Tan, black fin tips, paler below	Very common on coral reefs in knee-deep water with dorsal fins in air. Occurs in east Mediterranean, probably spreading from Red Sea via Suez Canal. Generally harmless but can be aggressive when bait is around. Has attacked many people, none seriously, but is regarded as a hazard.
Carcharhinus obscurus	Dusky shark, bay s, shovelnose, lazy-grey, black whaler	1,2,4,5,6,9,10,12,13,14,16. From surf zone to far offshore, all levels	400 ♀311 ♂310	Charcoal, copper-tinged, white below	Often follows ships. Has evil reputation but few attacks can definitely be attributed to it. The species most often taken in shark nets off South Africa. IGFA game fish.
Carcharhinus perezi	Caribbean reef shark	4,5,6. Bottom, inshore tropical waters	295 ♀247	Olive-grey, cream below	Abundant on Caribbean coral reefs. Dangerous, having attacked at least two people.
Carcharhinus plumbeus	Sandbar shark, brown s, ground s, northern whaler	4,5,6,8,9,10,11,12,13,14,16. Bottom, shallow inshore, deep offshore	239 ♀100 ♂177	Mouse-grey or copper, pale below	Abundant. Sometimes found in water barely covering it. Common in harbours and estuaries. Skin is esteemed for leather.
Carcharhinus porosus	Smalltail shark	2,3,4,5. Bottom, close inshore to moderately deep tropical waters	150 ♀109 ♂96	Slate, paler below	Common in estuaries. Harmless.
Carcharhinus sealei	Blackspot shark	10,11,13,14. Bottom, shallow, from surf zone to moderate depths	95 ♀81 ♂82	Mouse-grey, off-white below	Common. Too small to be considered dangerous.
Carcharhinus signatus	Night shark	4,5,6,9. Surface to deep, offshore and semi-oceanic waters	280	Slate, large green eyes, paler below	Probably migrates from depths to the surface at night.
Carcharhinus sorrah	Spot-tail shark	10,11,12,13,14,16. Shallow and moderately deep tropical waters	160 ♀130 ♂117	Grey, black fin tips, paler below	Common around coral reefs.
Carcharhinus wheeleri	Blacktail reef shark, short-nosed blacktail shark	10,11. Coral reefs, inshore and offshore waters	172 ♀146 ♂139	Charcoal, black/white fin tips	Aggressive near bait; should be considered potentially dangerous.
Galeocerdo cuvier	Tiger shark	1,2,3,4,5,6,7,9,10,11,12,13, 14,15,16. Worldwide in warm seas, surface and at moderate depths, close inshore and far offshore	740 ♀375 ♂298	Grey/black, tiger-like stripes fade with maturity	The shark of the Shark Arm Case. A maneater, second only to great white shark in number of known attacks on people and boats. Fearsome teeth and large mouth allow it to take almost anything: will eat sawsharks, hammerheads, turtles, sealions, jellyfish, lobsters and garbage. One giant weighed over 3 tonnes. Litters are between 10 and 82; at birth measures 51 to 76 cm (20 to 30 in). IGFA game fish.

Blacktip shark
Carcharhinus limbatus
225 cm

Oceanic whitetip shark
Carcharhinus longimanus
395 cm

Tiger shark
Galeocerdo cuvier
740 cm

Scientific name	Common names	Distribution*	Size (cm)*	Colour	Notes
Glyphis gangeticus	Ganges shark	11. Fresh water in rivers, estuaries, possibly also inshore waters	204	Brown to grey	Has horrific reputation in India, but the many deaths attributed to this fish were probably caused by the bull shark.
Glyphis glyphis	Speartooth shark	Possibly 13 and 14, probably inshore waters, possibly fresh water	100-300	Grey-brown	Until recently known from only one specimen (1839). More have been caught off Papua New Guinea and Queensland.
Isogomphodon oxyrhynchus	Daggernose shark	4. Tropical inshore waters and estuaries	152	Mustard-grey, paler below	The very long pointed snout and tiny eyes are adaptations to life in murky waters. Harmless.
Lamiopsis temmincki	Broadfin shark	11,13. Inshore waters	168 ♀146	Tan or ash	Little known. A slight potential danger because of large teeth.
Loxodon macrorhinus	Sliteye shark	10,11,12,13,14. Shallow inshore and moderately deep tropical waters	91 ♀85 ♂73	Grey, paler below	Has large eyes. Named 'sliteye' because of the little notches at the hind corners of the eyes.
Nasolamia velox	Whitenose shark, pico bianco	2,3. Moderately deep inshore and offshore tropical waters	150	Mouse-grey white-ringed dot on snout	Little known.
Negaprion acutidens	Sicklefin lemon shark, lemon s, sharp-toothed s	10,11,13,14,16. Surface and bottom, inshore waters, bays and reefs	310	Tawny, off-white below	Generally timid but will defend itself vigorously and obstinately if vexed, and is then dangerous. Thrives in aquariums.
Negaprion brevirostris	Lemon shark	2,3,4,5,6,9. Surface and at moderate depths, bays, near docks and in river mouths	340 ♀262 ♂251	Mustard/slate, fading below	Nega-prion = no saw, i.e. smooth teeth. Dangerous if provoked. Does so well in captivity it is widely used for research.
Prionace glauca	Blue shark, blue-dog, great blue s, blue whaler	Worldwide in tropical and temperate seas, often on surface of deep water	383 ♀276 ♂246	Brilliant blue back, white below	Found in wider range of waters than any other shark. Often seen swimming lazily with first dorsal fin and tip of tail out of water. In courtship males bite females and females develop skin three times as thick as males. Bears litters of up to 135; size when born between 35 and 45 cm (17.7 and 13.7 in). Extremely voracious; will gather round a food source in a frenzy. Packs cause havoc among hauls of fish, launching themselves at nets. Very dangerous; is known to have killed people and is suspected of causing great numbers of deaths after ship sinkings in World War II. IGFA game fish.
Rhizoprionodon acutus	Milk shark	9,10,11,12,13,14. Bottom, from shallow in surf zone to deep offshore waters	178 ♀117 ♂123	Grey above, white below	In India, flesh believed by some to improve the milk production of nursing mothers.
Rhizoprionodon lalandii	Brazilian sharpnose shark	4,5. Bottom, shallow and moderately deep tropical waters	77 ♀65 ♂54	Mouse-grey or charcoal, paler below	Common. Rhizo-prion-odon = teeth with serrated roots.
Rhizoprionodon longurio	Pacific sharpnose shark	1,2,3. Shallow, inshore tropical waters	128 ♀128 ♂75	Brown to grey, paler below	Very common.
Rhizoprionodon ologolinx	Grey sharpnose shark	11,13. Inshore and offshore tropical waters	70 ♀51 ♂45	Mouse-grey or grey, paler below	
Rhizoprionodon porosus	Caribbean sharpnose shark	4,5. Inshore and in deep estuaries, and deep offshore tropical waters	110 ♀94 ♂72	Metallic brown, paler below	Abundant. Taken in large numbers for food.

Daggernose shark
Isogomphodon oxyrhynchus
152 cm

Lemon shark
Negaprion brevirostris
340 cm

Blue shark
Prionace glauca
383 cm

*For note on size, and distribution map, see box p154.

Scientific name	Common names	Distribution*	Size (cm)*	Colour	Notes
Rhizoprionodon taylori	Australian sharpnose shark, Taylor's shark	14. Inshore tropical waters	67	Mouse-grey, abruptly white below	Common but little known.
Rhizoprionodon terraenovae	Atlantic sharpnose shark	5,6. Shallow, near surf zone, in harbours, bays and estuaries	110 ♀97 ♂84	Mouse-grey, white spots, paler below	Enters rivers but does not go far up them. Harmless and abundant.
Scoliodon laticaudus	Spadenose shark	10,11,12,13. Close inshore, lower reaches of rivers, tropical waters	74 ♀51 ♂41	Copper-tinged grey, pale below	Abundant in India and Pakistan. Often found in lower reaches of rivers, but length of survival in fresh water not known.
Triaenodon obesus	Whitetip reef shark, blunthead s	10,11,12,13,14,16. Bottom, shallow, in coral caves in reefs and lagoons	213 ♀131 ♂136	Brown/grey, white-tipped dorsal fin	Able to rest on the bottom. Timid, but can become excited by bait such as speared fish, and will bite humans. IGFA game fish.

Sphyrnidae (Bonnethead sharks, hammerhead sharks, scoophead sharks)

Scientific name	Common names	Distribution*	Size (cm)*	Colour	Notes
Eusphyra blochii	Winghead shark	11,12,13,14. Shallow, tropical waters	152	Mouse-grey, paler below	Has widest head of all the hammerheads, width sometimes half the length of body. Lobes are possibly to increase ability to find and take prey since the smelling and seeing organs are on them. Harmless.
Sphyrna corona	Scalloped bonnet-head	2,3. Probably inshore waters	92	Ashy, white below	The smallest of the hammerheads, its head is rounded and mallet-shaped. Appears scalloped when viewed from above.
Sphyrna couardi	Whitefin hammerhead	9. Bottom, coastal and pelagic, tropical waters	300 ♀232 ♂162	Slate to mouse-grey, pale below	Uncommon. Not known to attack people but its size makes it possibly dangerous.
Sphyrna lewini	Scalloped hammerhead, kidney-headed shark, bronze hammerhead	1,2,3,4,5,6,9,10,11,12,13,14, 16. All levels, tropical and warm waters, inshore and far offshore	420 ♀260 ♂214	Olive-brown, charcoal tips to fins	Most common hammerhead. Probably unaggressive. In some areas, forms huge schools when migrating. Has reputation for attacking people but the large hammerheads are often mistaken for one another. Scientists studying these hammerheads have found them unaggressive.
Sphyrna media	Scoophead	2,3,4,5. Tropical inshore waters	150 ♀116	Grey or mouse-grey, pale below	Sphyrna = hammer. All hammerheads are IGFA game fish.
Sphyrna mokarran	Great hammerhead	2,3,4,5,6,8,9,10,11,12,13, 14,16. Shallow reefs, and at moderate depths offshore	610 ♀365 ♂285	Dark olive, paler below	Sometimes found in water 1 m (3 ft) deep. Diet includes stingrays; these sharks have been found with many stings stuck in their mouths (one had about 50). Feared as a man-eater but is merely a strong suspect. Potentially very dangerous.
Sphyrna tiburo	Bonnethead, shovelhead, bonnet shark	1,2,3,4,5,6. Estuaries, bays and coral reefs, usually shallow waters	150 ♀107 ♂88	Mouse-grey, paler below	Head is quite rounded. Usually found in small groups. Tiburo = Spanish for shark. Closely studied by scientists who have discovered they have a wide range of 'body language' postures.
Sphyrna tudes	Smalleye hammerhead	4. Inshore waters	150 ♀134 ♂127	Mouse-grey, paler below	Known to eat young scalloped hammerheads, and also to enter fresh water. Tudes = hammer (Latin).
Sphyrna zygaena	Smooth hammerhead, balance fish, black hammerhead	Worldwide. Inshore and offshore, sub-tropical and tropical waters	400 ♀257 ♂233	Olive-grey, dusky tips to fins	Often seen at surface with dorsal fins and tail exposed. Considered dangerous. Young 55 cm (21.7 in) long at birth, in litters of 29 to 37.

Whitetip reef shark
Triaenodon obesus
213 cm

Winghead shark
Eusphyra blochii
152 cm

Scalloped hammerhead shark
Sphyrna lewini
420 cm

179

FACTS, FALLACIES AND RECORDS

Shark records – especially any claiming great size – are notoriously difficult to verify. The figures quoted here are those considered by most experts to be accurate. They err on the side of caution. It is almost certain that larger and heavier sharks have been seen or caught, but not enough evidence is available to confirm them as records.

The following questions are those most often asked about sharks. Not surprisingly, almost all are concerned with the dozen or so species that are popularly believed to be maneaters – the most 'shark-like' of this diverse and fascinating group of creatures.

Which is the largest shark?

The whale shark *Rhiniodon typus* is not only the world's largest shark, but also the world's largest fish. This plankton-eating giant reaches a length of perhaps 18 m (59 ft). The largest accurately measured specimen, caught off Pakistan in 1949, measured 12.65 m (41.5 ft) long and weighed 21.5 tonnes. There are numerous, almost certainly exaggerated, reports of specimens measuring up to 21 m (70 ft) long.

Next largest, after the whale shark, is the basking shark *Cetorhinus maximus,* also a filter feeder. Data on the sizes of these creatures is more reliable because they have been heavily fished by commercial organisations around the world for many years.

The largest accurately measured specimen reached 12.3 m (40.25 ft) and weighed an estimated 16 tonnes. It was caught off New Brunswick, Canada in August 1851. There are numerous claims made for record specimens which range up to an unlikely 15 m (50 ft), but none have been confirmed.

Which is the smallest shark?

The spined pygmy shark *Squaliolus laticaudus,* which grows to only about 250 mm (9.8 in), is probably the world's smallest shark. Males mature at around 150 mm (5.9 in) and females at 180 mm (7.1 in). A close contender for this title is the pygmy ribbontail catshark *Eridacnis radcliffei* which grows to about the same size. Males of this species mature at 180 mm (7.1 in) and females at 160 mm (6.3 in).

Alf Dean, facing the camera, and a large great white shark. Dean has held the world record for the largest shark ever caught on a fishing line since 1959. His record fish, which weighed 1208.38 kg (2664 lb), was caught on 60-kg (132.27-lb) breaking strain line.

Which is the largest maneater?

Few things in the popular world of sharks and shark stories are as much disputed as the record for the largest dangerous shark to be seen or caught. Considerable fame accompanies any claim to be associated with the largest maneater, so it is hardly surprising that individuals are tempted to exaggerate sightings and encounters. One important point to bear in mind is that the sizes and weights of sharks do not necessarily increase in proportion to one another. Much depends on the sex and condition of the fish.

It is also important to

A thresher shark *Alopias vulpinus* is hauled alongside a fishing boat off the coast of New Zealand. The tails on these extraordinary sharks are about the same length as their bodies. Threshers are thought to be harmless to humans, although there is a story of a fisherman decapitated by the tail of a large thresher in the Atlantic.

Not all shark fishermen use conventional means to secure their catches. This 1470-kg (3234-lb) great white shark was lassoed near Albany by West Australian Fisheries and Wildlife Officer Colin Ostle.

remember that it is often difficult to estimate the lengths of objects in the water, especially when their extremities are submerged or ill defined. In a revealing series of experiments, conducted at the Mote Marine Laboratory in Florida, students were asked to look at two sharks in a pool for 10 seconds, and then to estimate their lengths. Estimates for the length of a 1.8-m (72-in) bull shark ranged from 1.6 m (63 in) to 3.3 m (128 in), with a mean of 2.2 m (85.5 in). For a 1.2-m (46-in) lemon shark the estimates ranged from 74 cm (29 in) to 1.6 m (61 in), with a surprisingly low mean of 1.16 m (45.5 in).

The great white *Carcharodon carcharias* is the largest shark, after the big filter feeders. Its awesome reputation, and the fact that it has been responsible for more human fatalities than any other species, have meant that many myths have grown up around it over the years. It is now difficult to separate fact from fantasy in the stories that are told about it.

The largest great white shark actually weighed appears to be a 3312-kg (7302-lb) specimen caught off the coast of Cuba in 1945. It was 6.4 m (21 ft) long.

The largest great white actually measured was probably a monster caught by fishermen off the Azores in 1978. It was brought ashore and measured by a reliable observer at 9 m (29.5 ft). The giant fish was estimated to weigh around 4540 kg (10 000 lb).

Larger sharks have been claimed, but in every case it is difficult or impossible to verify the facts. There is a second hand report of a 11.3-m (37-ft) great white stranded in a weir at New Brunswick, Canada in the 1930s, but it has not been satisfactorily confirmed. Many claims have been made by sailors and fishermen for record sharks – especially in cases where it has been possible to measure a shark in the water alongside a boat of known length.

One claim often put forward concerns a great white caught off Port Fairy, Victoria in 1852. This specimen, it was claimed, measured 11.13 m (36.5 ft). However, the jaws of this shark are in the British Museum in London where they have been examined by researchers. Comparisons with jaws from sharks of known length seem to indicate that the Port Fairy shark probably only measured about 5.4 m (17.7 ft).

The record for the largest great white (or, indeed, any shark) caught on a rod and line – a claim that is recognised by the International Game Fishing Association – belongs to Alf Dean. His 1208.38-kg (2664.02-lb) monster was landed at Ceduna, South Australia, in 1959. It measured 5.13 m (16.83 ft). A 1566-kg (3452-lb) great white caught by Donald Braddock and Frank Mundus in August 1986 off the US east coast has yet to be verified by the IGFA. A 1537-kg (3388-lb) shark caught by Clive Green off Albany, Western Australia, in 1976 is not recognised by the IGFA because whale meat was used as bait, which is expressly forbidden under game fishing rules.

Next largest of the dangerous sharks is the tiger *Galeocerdo cuvier*. The largest accurately measured specimen, which was caught in a shark net off Newcastle, NSW in 1954, measured 5.5 m (18 ft) and weighed 1524 kg (3360 lb). Another shark of the same size was caught off Mackay, Queensland in 1980. Again, there have been exaggerated claims of tigers measuring 9 m (30 ft) or more.

The all-tackle gamefishing

The goblin, perhaps the strangest of all the world's sharks. Specimens of this deep-water species are rarely seen. Most of those captured are taken in deepwater trawls, or on longlines. Freshly caught goblin sharks are a pinkish-white colour.

record for a tiger stands at 807.4 kg (1780 lb), for a shark caught in 1964 off South Carolina, USA, by Walter Maxwell. This giant specimen was 4.229 m (13.88 ft) long.

Two other large shark species are the Greenland shark *Somniosus microcephalus* and the related Pacific sleeper shark *Somniosus pacificus*. In 1895 a 6.4-m (21-ft) Greenland shark was caught in the Firth of Forth, Scotland. It weighed 1021 kg (2250 lb). In 1966 two scientists in a research submarine at a depth of 1219 m (4000 ft) off the coast of California came upon a giant Pacific sleeper shark that they estimated to be 9.1 m (30 ft) long. Both of these sightings were of exceptionally large examples of their species.

Not all sharks are large
It is an interesting fact that of the 344 known species of sharks, only 39 are known to grow to over 3 m

(9.8 ft). Over half – 176 species – are under one metre (39 in) long at their maximum.

Which is the most common shark?
Piked dogfish *Squalus acanthias*, which are found in waters around the world from the coasts of Scandinavia to southern Chile, are probably the most common sharks. At various times they have supported important fisheries in several countries. It has been estimated that in 1904-05 some 27 million were taken off the coast of Massachusetts, USA, alone. On occasions longline fishermen have reported catching a piked dogfish on nearly every hook on a 1500-hook line. These sharks grow to a maximum size of only about 1.6 m (63 in).

Which is the rarest shark?
Fourteen sharks are known from one specimen only, but in most cases

these are fish that have a limited range, and perhaps live at great depths in the ocean.

It is a remarkable fact that a species of shark that can grow up to 4.5 m (14.8 ft) escaped detection until 1976. Only two specimens of the large, filter-feeding megamouth shark *Megachasma pelagios* have so far been caught. The first was discovered in strange circumstances when an oceanographic research vessel hauled in two parachute sea-anchors from a depth of about 165 m (541 ft) in waters off Hawaii, to find a megamouth entangled in one of them. It was suggested that the shark had been swimming along with its mouth open to catch small crustaceans when it encountered the parachute, which became entangled in its tiny teeth.

Which is the strangest shark?
This title must belong to any one of the hammerhead sharks (*Eusphyra blochii* or *Sphyrna sp.*) or to the bizarre, deep-water goblin shark *Mitsukurina owstoni*.

No really satisfactory explanation has been put forward for the strange shape of the hammerhead's head. It has been suggested that the wide separation of the fish's nostrils, and its habit of swinging its head from side to side while swimming, may enable it to

detect smells over a much wider area than would otherwise be possible. Other theories suggest that the shape of the head may make it a more efficient swimmer, with better manoeuverability.

The first specimen of a goblin shark was captured in waters off Japan in 1898. Fossil teeth from this strange shark had been known since the middle of the nineteenth century, but researchers had assumed it was extinct. Goblin sharks live in the deep oceans and are therefore rarely seen. The strange projections on the tops of their heads are probably an adaptation for feeding, although almost nothing is known about the lives of these curious creatures.

Which shark is most dangerous to humans?
Of the 344 known species of sharks, only about 27 are known to have attacked people or boats. Another 12 species are suspected of attacking people, and a further 28 have the potential to be dangerous. The most dangerous sharks

(excluding the plankton eaters) are all over two metres (6.6 ft) long.

Of the sharks positively identified in attacks on people, three stand out as being most dangerous – the great white *Carcharodon carcharias,* the tiger *Galeocerdo cuvier* and the bull *Carcharhinus leucas.* Between them these three sharks have been responsible for at least 80 deaths. The most blamed was the great white with 32 attacks, next was the tiger with 27 attacks and last the bull with 21 attacks.

How long do sharks live?
The life span of sharks has been something of a mystery for many years. Only recently, as the results of tagging studies start to become available (see p 47), has it been possible to make realistic estimates. Generally, it appears that the slow-growing sharks live longest. Both the Port Jackson shark *Heterodontus portusjacksoni* and the whitetip reef shark *Triaenodon obesus* probably live to about 25 years, while the tope shark

---

REAPING THE BENEFITS OF FRIENDSHIP

The remarkable sucking disc on the head of a remora is difficult to dislodge once attached.

Sharks are often seen with two companions – pilot fish and remoras. The relationship between pilot fish and sharks is little understood. They do not appear to depend on sharks for food, nor do they lead sharks, as was once suggested.

Remoras have a disc on the tops of their

heads with which they can attach themselves to the skin of a shark, or any other large fish or floating object, by vacuum action. Remoras, like pilot fish, do not generally share a shark's food – the only benefit they receive from the association seems to be free travel.

Galeorhinus galeus and the piked dogfish *Squalus acanthias* may have a life span of around 40 years. Scientists have claimed that some piked dogfish may live as long as 100 years, and a similar figure is sometimes suggested for large great white sharks *Carcharodon carcharias*.

How fast do sharks swim?

Most sharks are slow swimmers, only reaching maximum speed while chasing food. Few actual measurements have been made, but the work that has been done suggests that the warm blooded species, such as the mako *Isurus oxyrhinchus*, porbeagle *Lamna nasus* and great white *Carcharodon carcharias* can all sustain higher swimming speeds than other sharks. On occasions makos will leap from the water, and in order to do that they must reach at least 35 km/h (22 mph) in short bursts. Other researchers claim to have measured steady speeds of 39 km/h (24.5 mph) from blue sharks *Prionace glauca,* with short bursts of 69 km/h (43 mph), although this seems unlikely.

An idea of normal cruising speeds can be gained from tagging studies of sharks during migrations. The fastest sustained trip so far recorded was by a mako which covered 2413 km (1500 miles) in 86 days at a rate of 28 km (17.6 miles) a day. Shorter journeys of 400 km (248 miles), or more, in seven days are common. This means the journeys must have been accomplished at speeds of at least 2.4 km/h (1.5 mph). However, because sharks do not normally swim in straight lines all the time, they would have covered considerably more than 400 km during such a journey. During normal day-to-day activities sharks are thought to swim around at a much more sedate pace of about 1 km/h (0.7 mph).

How deep do sharks go?

So little is known about the great ocean depths that it is impossible to give an accurate answer to this question. A number of deep water sharks have been caught below 3500 m (11 500 ft), but they may easily descend to the deep ocean floor on occasions (the deepest ocean trench plunges to 11 038 m [36 204 ft] below sea level).

Some deep water species, such as the cookiecutter shark *Isistius brasiliensis,* rise to the surface to feed at night, making a remarkable journey which involves a vertical swim of around 7 km (4.4 miles) there and back.

Which shark is the greatest traveller?

This distinction undoubtably belongs to the blue shark *Prionace glauca*. Migrations of from 2000 to 3000 km (1200 to 1700 miles) are common, and the record so far is for a journey of 5980 km (3740 miles) from New York State to Brazil.

Other sharks are also great travellers, and records exist of movements of over 2500 km (1550 miles) by makos *Isurus oxyrhinchus,* tigers *Galeocerdo cuvier* and sandbar sharks *Carcharhinus plumbeus.* Bull sharks *Carcharhinus leucas* have been found over 3700 km (2300 miles) from the sea in the upper reaches of the Amazon, so they are presumably capable of swimming that distance.

Do sharks live in fresh water?

Only two species of sharks are conclusively proved to venture into fresh water – the bull shark *Carcharhinus leucas* and the Ganges shark *Glyphis gangeticus,* although half a dozen others are suspected of entering streams, and several can tolerate brackish water for short periods of time.

At one time bull sharks caused considerable confusion. They are found in one major lake, Lake Nicaragua, and many rivers, including the Ganges, the Zambezi, the Mississippi, the Tigris and the Amazon. Researchers thought that several of these populations were in fact different species, and named them accordingly. Only in recent years has it become obvious that they are all bull sharks.

Bull sharks have also been confused with the only other proven freshwater species, the Ganges shark. This is an extremely rare fish which is known only from a few specimens taken from the Ganges and Hooghly Rivers in India. It has acquired a fearsome reputation as a maneater, but this may be undeserved; the real culprit may be the bull shark.

Is it true that sharks must turn on their backs to attack?

It is often asserted that sharks cannot bite while swimming the right way up, which is completely incorrect. This idea was first put forward by Aristotle in 330 BC in his *Historia Animalium,* and is still believed by many people today. Sharks can, and do, attack their prey in whichever way is most convenient.

Are sharks attracted by human blood?

This is another 'truth' that is frequently quoted. In fact experiments have shown that sharks do not appear to be particularly interested in the blood of any mammal, although they are very sensitive to extremely small concentrations of fish blood in water. Further confirmation of this fact is provided by the number of cases in which sharks have bitten humans once and then lost interest. Despite this, some experts still suggest that anyone with an open wound would do well to avoid swimming in water where sharks may be found – just in case.

WORLD GAME FISHING RECORDS FOR SHARKS

John Robinson, *an Australian angler, once held the world record with this 620-kg (1422-lb) tiger shark. The title is now held by an angler in the USA.*

Broadly, the aim of game fishing is to catch the largest possible fish on the lightest possible line. Ten weights of line may be used, ranging from 1 kg (2.2 lb) breaking strain to 60 kg (132.27 lb) breaking strain. Records are established for the heaviest fish of each species caught on the various line weights. Local records are verified by the various national associations, and world records by the International Game Fish Association in the United States of America.

Strict rules govern the sport, the expressed aim being to give the fish a fighting chance. Lines, hooks, rods and other equipment must comply with a rigid set of definitions. Only the angler who hooks the fish is allowed to play it – no other person is allowed to help, or to touch the rod or line.

The following records are for 'all tackle' – any weight of line up to the maximum of 60 kg – and were those recognised by the International Game Fishing Association in mid-1986.

Bob Dyer *and a great white shark weighing over 900 kg (2000 lb). Dyer died in 1984, but many of his fishing records, for various line weights, still stand today.*

Species	Weight	Length	Place caught	Date	Angler
Blacknose shark	13.60 kg 30.00 lb	114.90 cm 45.25 in	Neptune Beach, Florida, USA	21 Jul. 1985	Mr J. H. David
Blue shark	198.22 kg 437.00 lb	363.22 cm 143.00 in	Catherine Bay, NSW, Australia	2 Oct. 1976	Mr P. Hyde
Bull shark	220.44 kg 486.00 lb	— —	Key West, Florida, USA	10 Apr. 1978	Mr P. Peacock
Dusky shark	346.54 kg 764.00 lb	297.18 cm 117.00 in	Longboat Key, Florida, USA	28 May 1982	Mr W. Girle
Greenland shark	432.00 kg 952.38 lb	— —	Trondheimsfiord, Norway	19 May 1984	Mr E. Nielsen
Hammerhead shark	449.50 kg 991.00 lb	447.04 cm 176.00 in	Sarasota, Florida, USA	30 May 1982	Mr A. Ogle
Lemon shark	180.07 kg 397.00 lb	304.80 cm 120.00 in	Dunedin, Florida, USA	29 Apr. 1977	Mr R. M. Guccione
Mako shark	489.88 kg 1080.00 lb	345.44 cm 136.00 in	Montauk, New York, USA	26 Aug. 1979	Mr J. L. Melanson
Porbeagle shark	210.92 kg 465.00 lb	281.94 cm 111.00 in	Padstow, Cornwall, England	23 Jul. 1976	Mr J. Potier
Sandtiger shark	144.24 kg 318.00 lb	251.46 cm 99.00 in	Nags Head, North Carolina, USA	25 May 1986	Mr D. Wolfe
Sixgill shark	164.50 kg 362.63 lb	250.00 cm 98.43 in	S. Miguel, Azores Islands	14 Aug. 1985	Mr F. Schopf
Spinner shark	40.48 kg 89.25 lb	148.59 cm 58.50 in	Isla Coiba, Panama	22 Aug. 1979	Mr R. Vrablik
Spiny dogfish	3.62 kg 8.00 lb	83.19 cm 32.75 in	Cape Cod Bay, Massachusetts, USA	26 Aug. 1977	Mr D. E. Singer
Thresher shark	363.80 kg 802.00 lb	487.98 cm 192.12 in	Tutukaka, New Zealand	8 Feb. 1981	Ms D. North
Tiger shark	807.40 kg 1780.00 lb	422.91 cm 166.50 in	Cherry Grove, South Carolina, USA	14 Jun. 1964	Mr W. Maxwell
Tope shark	32.49 kg 71.63 lb	— —	Knysna, Republic of S. Africa	18 Jan. 1984	Mr W. F. De Wet
White shark	1208.38 kg 2664.00 lb	513.08 cm 202.00 in	Ceduna, South Australia	21 Apr. 1959	Mr A. Dean
Whitetip reef shark	18.25 kg 40.25 lb	121.92 cm 48.00 in	Isla Coiba, Panama	8 Aug. 1979	Mr J. Kamerman

NORTH AMERICA

Attacks on the east coast of the United States are fairly well distributed between New York and Miami. On the west coat they are concentrated around San Francisco and Los Angeles. There are surprisingly few attacks in Mexico, but this may just be because data are scarce. Only a handful of attacks have ever occurred in Canadian waters. (For details see pages 194-95.)

GREAT BRITAIN

Scotland can boast the most northerly (provoked) attack in the world – off Wick in 1960. Several species of sharks, including dangerous ones, are seen near British shores. (For details see pages 196-97.)

THE CARIBBEAN SEA

The warm waters of the Caribbean are home to many shark species and this is reflected in the large number of attacks that occur along the coasts of Cuba, Haiti and northern South America. (For details see pages 196-97.)

PACIFIC ISLANDS AND NEW ZEALAND

The greatest concentrations of attacks have been around Papua New Guinea, among the islands of the Bismarck Archipelago (to the north of New Guinea) and in Hawaii. There are surprisingly few attacks recorded from other Pacific Islands where the inhabitants regularly use the ocean, but this may be because of problems in obtaining reliable reports of incidents. New Zealand has recorded many attacks, including the world's most southerly. (For details see pages 190-91).

SOUTH AMERICA

Surprisingly few attacks are reported from any part of South America below the equator. It seems likely that this is because there have been problems in collecting data from these areas. (For details see pages 196-97.)

SEE MAP P. 194-95
SEE MAP P. 190-91

GLOBAL PATTERNS OF SHARK ATTACK

The following pages examine, with maps and graphs, shark attacks around the world. Sites of almost all known recorded attacks are pinpointed, and attack figures are broken down month by month, to see if any seasonal patterns emerge.

In 1958 the United States Navy provided funds for a Shark Attack File to be established at the Smithsonian Institution in Washington, DC. Using old records, reports from scientists, divers and news clipping services, the project compiled a file of 1652 attacks from around the world before funding ceased in 1967. It remains the most detailed list of attacks available.

The information was analysed by Dr H. David Baldridge at the Mote Marine Laboratory in Florida. In assessing the reliability of the information on the File, he noted that in 90 per cent of cases accounts were based largely on information provided by people who were not at the scene of the attack. He also pointed to a geographic or cultural bias in the File – cases came predominantly from English-speaking countries. This presumably reflected a language barrier, rather than any scarcity of attacks in other regions.

Despite these problems, David Baldridge's analysis provides the most reliable picture of shark attack patterns that exists. It also overturns some long and widely held beliefs about shark attacks and shark behaviour.

Over two-thirds of all the documented attacks have occurred since 1940. This is an average of about 28 cases a year, world wide, (less than the 100 cases often cited as a likely average).

All but four of the attacks in the File occurred between 47°S and 46°N. Generally there were very few attacks near the equator, rising to a peak in the middle latitudes and falling off again in colder waters. The correlation between

THE MEDITERRANEAN SEA
Many species of sharks, including some dangerous to humans, are found in the Mediterranean. Several attacks have been recorded in the Adriatic Sea, off the coasts of Italy and Yugoslavia, including the most northerly recorded unprovoked attack in 1934. (For details see pages 196-97.)

EAST ASIA
There is a small concentration of attacks around Singapore, and a few have been recorded off the coasts of China and Japan, but less than might be expected. (For details see pages 196-97).

SOUTH ASIA
The Ganges delta and the Persian Gulf are both areas notorious for shark attacks. However, reports of many attacks on these coasts probably do not reach researchers. (For details see pages 196-97.)

AUSTRALIA
Australia has recorded more attacks than any other country on earth. The majority are concentrated around the major cities, and the others are distributed in a way that reflects the large number of people who live on the coastal fringe. (For details see pages 188-89.)

AFRICA
The most remarkable feature of African shark attacks – apart from those that take place in South Africa – is that there appear to be so few of them. This is almost certainly due to the fact that most occur in remote areas and are not reported. (For details see pages 192-93.)

P. 192

SEE MAP P. 193

SOUTH AFRICA
Most attacks in South Africa occur around Durban, on the Natal coast. A few take place around Cape Town, but none have been recorded on the Atlantic coast north of that city. (For details see pages 192-93.)

SEE MAP P. 188-89

SEE MAP P. 190-91

water temperature and shark attack, however, has more to do with the likelihood of people being in the water when it is warm, than with sharks preferring that temperature. The 'shark attack season' in any locality would be the time when the water was warm enough for humans (above 20°C; 68°F) and not too hot for sharks (less than 30°C; 86°F).

Only a third of recorded attacks were fatal, and the number of fatal attacks is declining. This decline could be due to the greater likelihood today of a victim receiving prompt medical attention.

Baldridge's most significant conclusions concern shark

motivation. Only one-quarter of all attack victims received the types of wounds that suggested that the shark's main interest was in feeding. Only 20 per cent of all attacks involved more than one strike, and in only four per cent of cases did the shark behave in a frenzied manner. None of these findings indicate that sharks, which are extremely efficient marine predators, are particularly interested in eating humans.

Baldridge also found no evidence for the popular theory that wounded or bleeding people were more likely to be attacked, or that sharks were attracted to any mammalian blood.

The attacks plotted on the

following maps and graphs are taken from a list of approximately 1200 attacks that were in the Shark Attack File in 1967. They show unprovoked and provoked attacks – the latter being where a shark was caught, trapped, speared or somehow aggravated before it eventually attacked.

Attacks on boats, on victims of sea and air disasters (where there is no way of knowing whether the victims were alive or not before the attack) and other doubtful attacks have been excluded from the maps and graphs. The numbers beside some of the markers indicate the number of attacks at that location, and not the number of victims.

AUSTRALIAN SHARK ATTACKS

The earliest known attack in Australian waters was in 1803 at Hamelin Bay in Western Australia, when a Mr Lefevre was attacked by a shark and died from his wounds. The unfortunate man was the first of 319 recorded Australian shark attack victims – over one-quarter of all the unprovoked and provoked attacks from around the world recorded in the Shark Attack File.

The distribution of attacks around the coastline largely reflects population densities and the popularity of water activities. With about 220 attacks, the east coast of Australia might, on the face of it, appear to be one of the most shark infested areas on earth, but this high number really only reflects the fact that there are a large number of beaches on this coast, with many people using them.

Sydney beaches were notorious for attacks until meshing was introduced in the 1930s. Since then there has not been a shark fatality at a Sydney ocean beach (up to the end of 1985), although attacks still occur within the harbour.

The seas off southern Australia are one of the main haunts of the great white shark, *Carcharodon carcharias*, one of the most dangerous of all sharks. Many of the shark sequences for the film *Jaws* were shot there, and nearly all the record-breaking great white sharks have been caught in these waters. After California, the southern coast of Australia has the second highest incidence of great white shark attacks.

For many years Bass Strait was thought to be the southern limit of shark attacks in Australia. But in January 1959 a naval rating was attacked and killed several hundred metres off Port Arthur in the south-east of Tasmania. Five days later another attack occurred off Port Davey in the south-west. They remain the most southerly attacks in Australian waters.

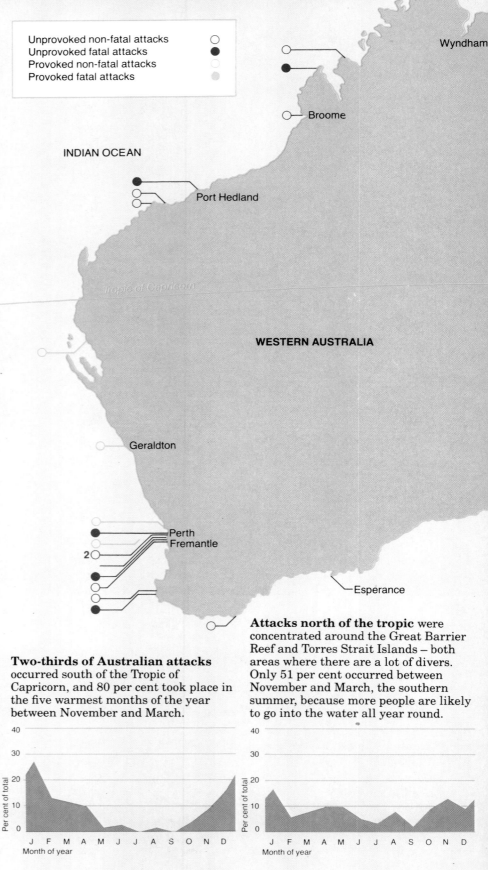

Unprovoked non-fatal attacks ○
Unprovoked fatal attacks ●
Provoked non-fatal attacks ◦
Provoked fatal attacks ◦

Two-thirds of Australian attacks occurred south of the Tropic of Capricorn, and 80 per cent took place in the five warmest months of the year between November and March.

Attacks north of the tropic were concentrated around the Great Barrier Reef and Torres Strait Islands – both areas where there are a lot of divers. Only 51 per cent occurred between November and March, the southern summer, because more people are likely to go into the water all year round.

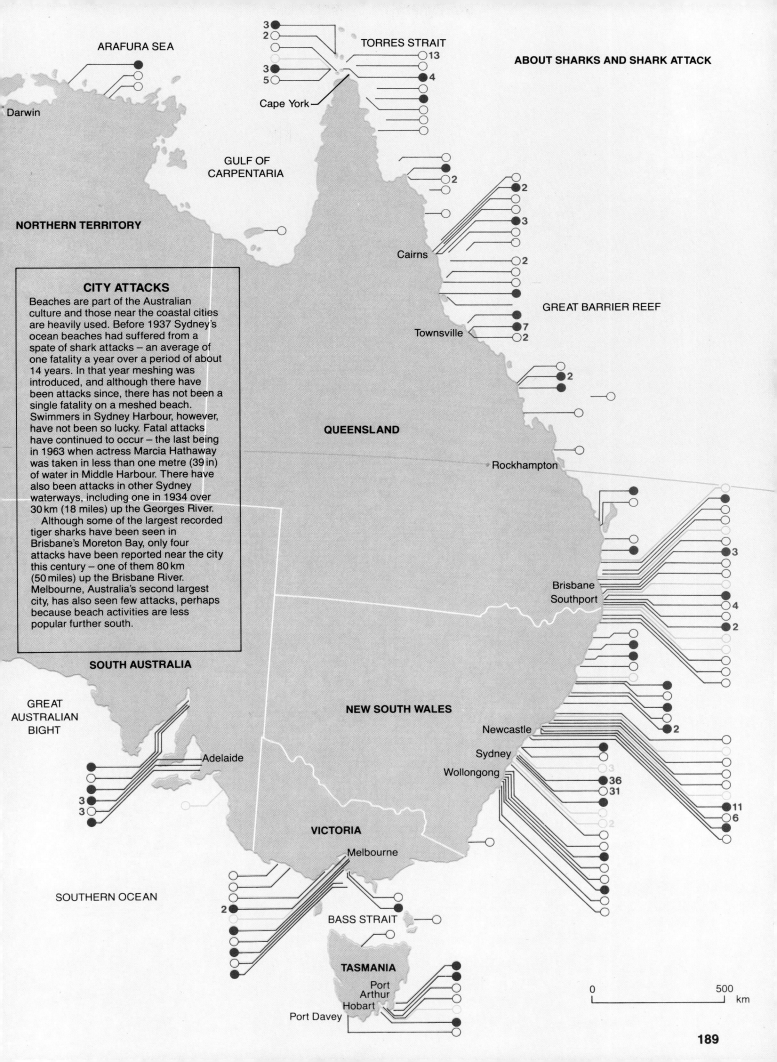

ARAFURA SEA

Darwin

TORRES STRAIT

ABOUT SHARKS AND SHARK ATTACK

3
2

13

3
5

4

Cape York

GULF OF
CARPENTARIA

NORTHERN TERRITORY

2

2
2

Cairns

3

2

GREAT BARRIER REEF

Townsville

7
2

2

QUEENSLAND

Rockhampton

CITY ATTACKS

Beaches are part of the Australian culture and those near the coastal cities are heavily used. Before 1937 Sydney's ocean beaches had suffered from a spate of shark attacks – an average of one fatality a year over a period of about 14 years. In that year meshing was introduced, and although there have been attacks since, there has not been a single fatality on a meshed beach. Swimmers in Sydney Harbour, however, have not been so lucky. Fatal attacks have continued to occur – the last being in 1963 when actress Marcia Hathaway was taken in less than one metre (39 in) of water in Middle Harbour. There have also been attacks in other Sydney waterways, including one in 1934 over 30 km (18 miles) up the Georges River.

Although some of the largest recorded tiger sharks have been seen in Brisbane's Moreton Bay, only four attacks have been reported near the city this century – one of them 80 km (50 miles) up the Brisbane River. Melbourne, Australia's second largest city, has also seen few attacks, perhaps because beach activities are less popular further south.

3

Brisbane
Southport

4

2

SOUTH AUSTRALIA

GREAT
AUSTRALIAN
BIGHT

Adelaide

3
3

NEW SOUTH WALES

2

Newcastle
Sydney
Wollongong

3
36
31
2

11
6

VICTORIA

Melbourne

SOUTHERN OCEAN

2

BASS STRAIT

TASMANIA

Port
Arthur
Hobart

Port Davey

0   500
         km

189

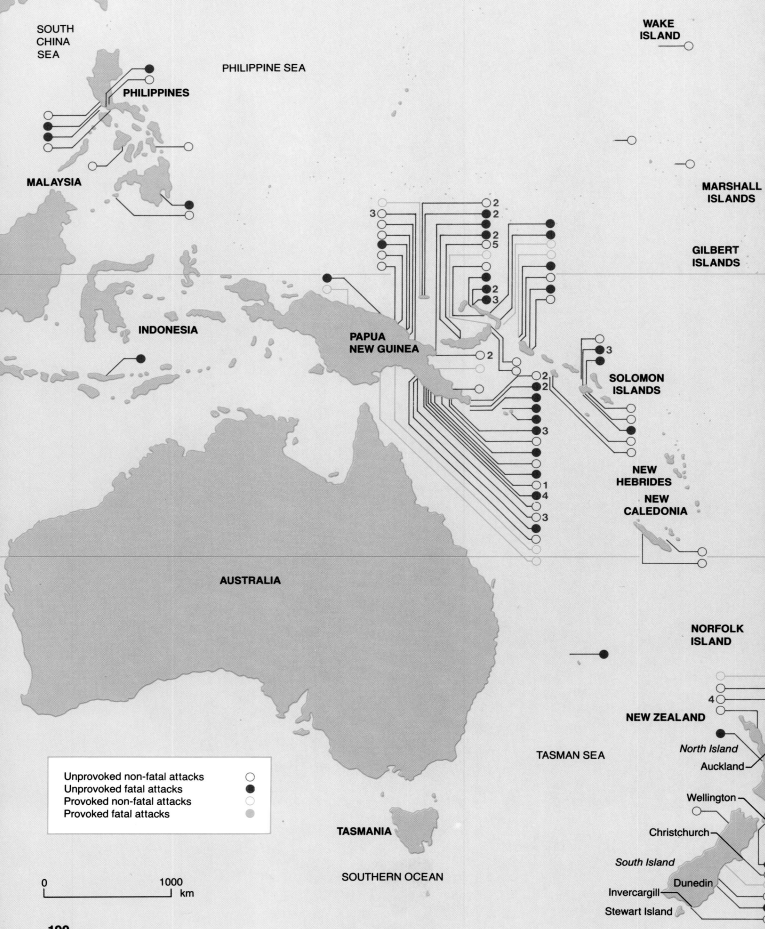

TAIWAN

SOUTH
CHINA
SEA

PHILIPPINE SEA

WAKE
ISLAND

PHILIPPINES

MALAYSIA

MARSHALL
ISLANDS

GILBERT
ISLANDS

3

2
2
2
5

2

INDONESIA

PAPUA
NEW GUINEA

2

2
2

3

2
2

3

3

SOLOMON
ISLANDS

NEW
HEBRIDES

NEW
CALEDONIA

1
4

3

AUSTRALIA

NORFOLK
ISLAND

4

NEW ZEALAND

North Island

Auckland

Wellington

TASMAN SEA

Christchurch

South Island

TASMANIA

Dunedin

Invercargill

SOUTHERN OCEAN

Stewart Island

Unprovoked non-fatal attacks ○
Unprovoked fatal attacks ●
Provoked non-fatal attacks ○
Provoked fatal attacks ●

0   1000
                        km

190

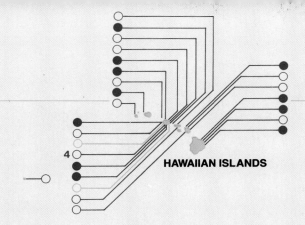

MIDWAY ISLANDS ○2

HAWAIIAN ISLANDS

4 ○

PACIFIC OCEAN

Pacific Ocean attacks show no obvious shark attack 'season', since nearly all the island groups are in the tropics where diving, swimming and fishing are year-round activities.

CHRISTMAS ISLAND

Equator

PHOENIX ISLANDS ○

TUVALU

TOKELAU ISLANDS

WESTERN SAMOA

AMERICAN SAMOA ●2

TONGA

○3

FIJI

COOK ISLANDS

FRENCH POLYNESIA

New Zealand attacks show a marked seasonal pattern. None were recorded at all in the months between April and November, reflecting the small number of people who are prepared to venture into cold southern waters at that time of the year.

Chart 1: Per cent of total / Month of year — J F M A M J J A S O N D (scale 0, 10, 20, 30, 40)

Chart 2: Per cent of total / Month of year — J F M A M J J A S O N D (scale 0, 10, 20, 30, 40)

Tropic of Capricorn

PITCAIRN GROUP

New Zealand And Pacific Ocean Attacks

The map shows the locations of 189 attacks that have been recorded for the Pacific area – 37 per cent of which were fatal – a proportion that is in line with the Shark Attack File as a whole.

Among the cultures of the Pacific islands the importance of sharks is reflected in folklore and myth, in shark gods and shark legends. It would have seemed likely that a large number of attacks should have been recorded from these islands, at least since regular records of dates and facts were kept

after western contact. However, isolation and distance have meant that probably only a fraction of the incidents have reached the Shark Attack File.

The area with the greatest concentration of recorded attacks is Hawaii. Most have been on fishermen – mainly of Japanese extraction – and ordinary swimmers seem to have been relatively immune.

One of the worst mass shark attacks of modern times took place with the sinking of the steamer

Negros off Bondoc Peninsula in the Philippines in 1927. Fifty-five people died, many of them having been taken by sharks as they floundered helpless in the water.

The world's most southerly recorded attack was on 27 January 1962 at 46°S, off the southern end of the South Island of New Zealand. Norman McEwan received deep gashes to his wrist from a 1.5 m (4.9 ft) shark while swimming in waist-deep water. More attacks, some fatal, have been recorded from other parts of the South Island.

AFRICAN AND SOUTH AFRICAN ATTACKS

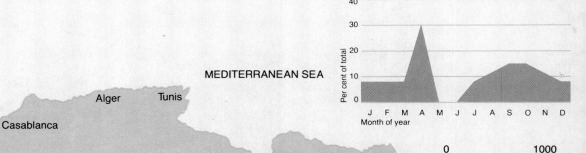

There is no obvious seasonal pattern to attacks over the greater part of Africa. However, these figures are far from complete, so no significant conclusions can be drawn from them.

Per cent of total

40
30
20
10
0

J F M A M J J A S O N D
Month of year

0   1000
         km

MEDITERRANEAN SEA

Alger   Tunis

Casablanca

Alexandria

Tropic of Cancer

RED SEA

Dakar

Conakry

Freetown

Monrovia   Lagos
    Accra
Abidjan   Port Harcourt

ATLANTIC OCEAN

Equator

Luanda

Djibouti

Mombasa

Dar-es-Salaam

Moçambique

Tropic of Capricorn   Walvisbaai

Maputo

Madagascar

INDIAN OCEAN

Cape Town

SOUTHERN OCEAN

The most obvious feature of the map of South African shark attacks is the number concentrated along the eastern coastline of that country. Few people swim along the Atlantic coast due to a cold current that sweeps in from the polar regions, and there have been no recorded attacks along this coast between Cape Town at 34°S and Monrovia, in Liberia, at 6°N. The currents along the eastern coast are warmer, however, and more suitable for water activities.

There is an over-representation of South African attacks which is almost certainly due to the fact that reports of attacks from there were more easily available to researchers compiling the Shark Attack File. It is also noticeable that the majority of recorded South African victims were white, although this group

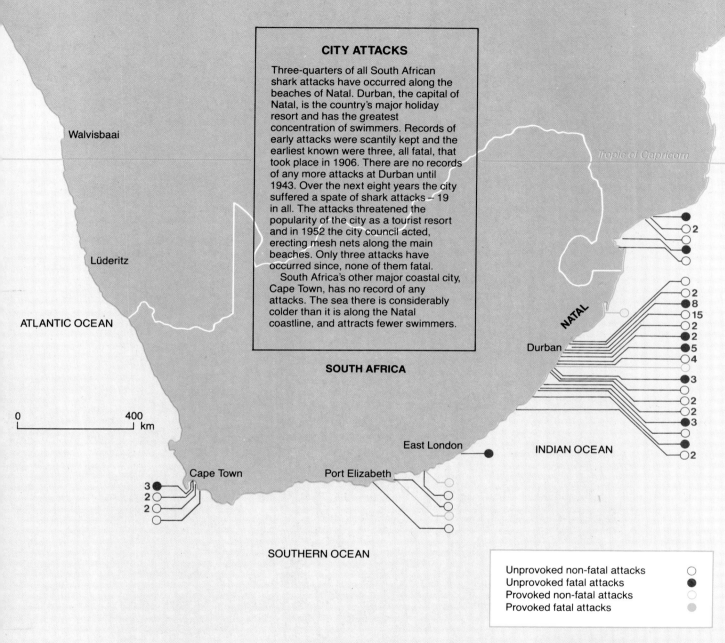

CITY ATTACKS

Three-quarters of all South African shark attacks have occurred along the beaches of Natal. Durban, the capital of Natal, is the country's major holiday resort and has the greatest concentration of swimmers. Records of early attacks were scantily kept and the earliest known were three, all fatal, that took place in 1906. There are no records of any more attacks at Durban until 1943. Over the next eight years the city suffered a spate of shark attacks – 19 in all. The attacks threatened the popularity of the city as a tourist resort and in 1952 the city council acted, erecting mesh nets along the main beaches. Only three attacks have occurred since, none of them fatal.

South Africa's other major coastal city, Cape Town, has no record of any attacks. The sea there is considerably colder than it is along the Natal coastline, and attracts fewer swimmers.

Legend:
- ○ Unprovoked non-fatal attacks
- ● Unprovoked fatal attacks
- ○ Provoked non-fatal attacks
- ○ Provoked fatal attacks

makes up less than 20 per cent of the population. It seems likely that this was because more whites take part in water sports, rather than a preference on the part of sharks for people with white skins.

Attacks on the beaches of Natal account for three-quarters of all South African cases. In 1940 a remarkable sequence of five fatal attacks took place just south of Durban, the capital of Natal.

Natal was also the location of one of the worst mass shark attacks on record. In 1942, off the northern coast, the *Nova Scotia*, carrying 1000 Italian prisoners of war, was torpedoed. Most people on board died, with a large, although unknown, number taken by sharks.

Another notorious incident, also involving a shipwreck, took place about 100 km (62 miles) to the east of Cape Town at Danger Point when a British troopship, the *Birkenhead*, went aground in 1852. Many of the 455 lives that were lost were due to attacks by packs of sharks.

South African attacks, as in other temperate regions of the world, show a strong seasonal pattern. Seventy-five per cent took place in the five warmer months from November to March, when many people go swimming.

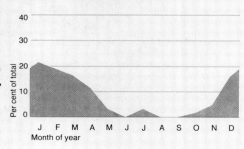

UNITED STATES AND CANADIAN ATTACKS

North America, after Australia, has the highest incidence of shark attack in the world: 253 unprovoked and provoked attacks have been recorded – 225 of them along both coasts of the United States, and the remainder in Mexico. Thirty-two per cent of all attacks were fatal. The most northerly (confirmed) attack was off Swampsott, Massachusetts (42° 28′N).

Of the United States attacks, three-quarters took place along the east coast. This coastline is longer than the west coast and the proportion is in line with the relative numbers of people living along the two coasts. The west coast, however, probably has the greatest concentration of great white sharks in the world. Almost 40 per cent of all great white shark attacks have occurred along a 200-km (124-mile) stretch of California.

One of the most notorious of all shark attack sagas occurred off the New Jersey coast in 1916. Over a period of 12 days, along a densely populated section of the coast, five people were attacked. The response was hysterical. Guns and explosives were used against anything that moved in nearby waters for a period of several weeks in an attempt to find the shark responsible. It was assumed that the attacks were the work of one rogue shark (although that theory is now in doubt).

Although sharks, including species dangerous to humans, are common in Canadian waters, there appear to have been few attacks. Author R. M. Ballentyne reported an incident that took place in 1848 in which an Indian family were attacked while canoeing in the Gulf of St. Lawrence. They only escaped by throwing a baby overboard to distract the shark. On another occasion a warden was menaced by a shark while he was crossing ice floes near Basque Island in 1940.

UNITED STATES OF AMERICA

3
2
2
2
San Francisco

3
3
3

Los Angeles

2
2
3
San Diego

PACIFIC OCEAN

Unprovoked non-fatal attacks	○
Unprovoked fatal attacks	●
Provoked non-fatal attacks	○
Provoked fatal attacks	○

0   500
|_____| km

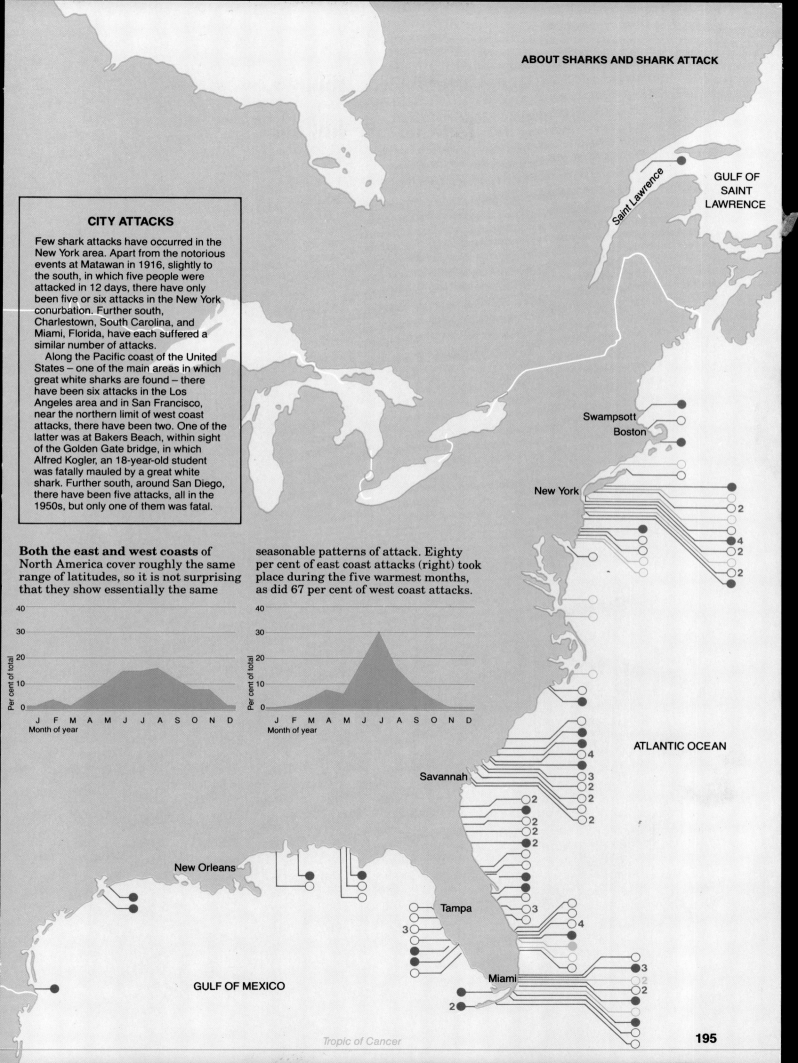

GULF OF
SAINT
LAWRENCE

Saint Lawrence

CITY ATTACKS

Few shark attacks have occurred in the
New York area. Apart from the notorious
events at Matawan in 1916, slightly to
the south, in which five people were
attacked in 12 days, there have only
been five or six attacks in the New York
conurbation. Further south,
Charlestown, South Carolina, and
Miami, Florida, have each suffered a
similar number of attacks.

Along the Pacific coast of the United
States – one of the main areas in which
great white sharks are found – there
have been six attacks in the Los
Angeles area and in San Francisco,
near the northern limit of west coast
attacks, there have been two. One of the
latter was at Bakers Beach, within sight
of the Golden Gate bridge, in which
Alfred Kogler, an 18-year-old student
was fatally mauled by a great white
shark. Further south, around San Diego,
there have been five attacks, all in the
1950s, but only one of them was fatal.

Swampsott
Boston

New York

Both the east and west coasts of
North America cover roughly the same
range of latitudes, so it is not surprising
that they show essentially the same

seasonable patterns of attack. Eighty
per cent of east coast attacks (right) took
place during the five warmest months,
as did 67 per cent of west coast attacks.

ATLANTIC OCEAN

Savannah

New Orleans

Tampa

Miami

GULF OF MEXICO

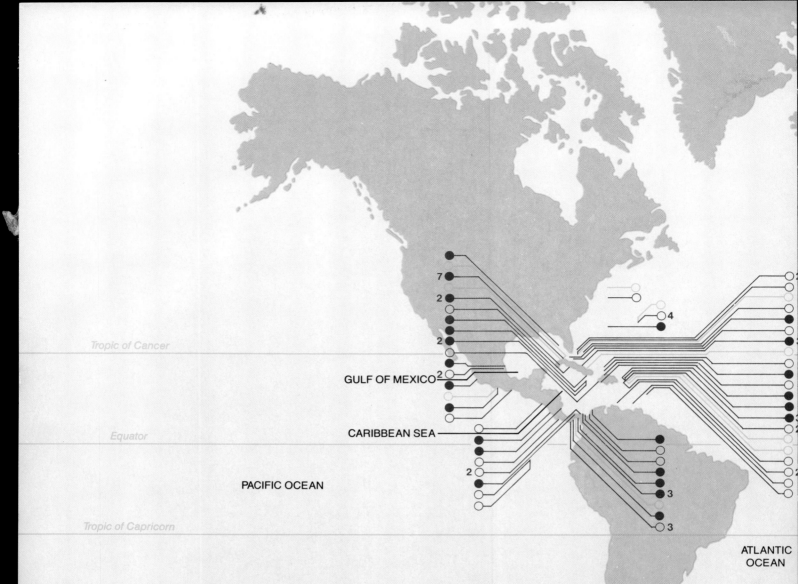

7
2
2
GULF OF MEXICO 2
CARIBBEAN SEA

Tropic of Cancer

Equator

PACIFIC OCEAN

Tropic of Capricorn

4

2

3

3

ATLANTIC
OCEAN

ATTACKS AROUND THE REST OF THE WORLD

This map shows the locations of shark attacks in areas that are not included in the preceding maps.

Europe has been relatively free of attacks, although a number have been recorded in the eastern Mediterranean, especially in the Adriatic Sea. The most northerly Mediterranean attack was in the Adriatic in 1934 when 18-year-old Agnes Novak was fatally injured near Susak, Jugoslavia.

The only other European cases outside the Mediterranean both involved sharks caught by fishermen. The most northerly of all attacks took place in the North Sea off Wick, Scotland (58° 26′N), in 1960, when Hans Schapper was bitten on the arm by a small shark

that had been dragged on board a trawler in a net full of fish.

Two-thirds of European attacks have been fatal – twice the world-wide average.

An area with a particularly bad shark attack record is the Persian Gulf. A number of fatal attacks have taken place in the pearl beds and fishing grounds at Ra's At-tannūrah, Saudi Arabia. The river complex at the head of the Gulf is also dangerous. Fatal attacks have occurred 150 km (93 miles) up the Karun River near Ahvāz in Iran.

Another area notorious for sharks is the Ganges Delta, which straddles the border between India and Bangladesh, especially the Hooghly River that flows through

Calcutta. In 1880 alone 20 people were attacked in this area. Further to the east, in Singapore, Hong Kong and China, there have been numerous attacks on divers, swimmers and fishermen.

It seems strange that a small number of attacks (four only) seem to have taken place in Japanese waters. This could be a result of language difficulties that have prevented attacks from being recorded by American researchers.

The same problem may occur in South America which has not apparently been the site of many shark attacks at all. The only recorded South American attack below the equator was at Buenos Aires in 1954. It was not fatal.

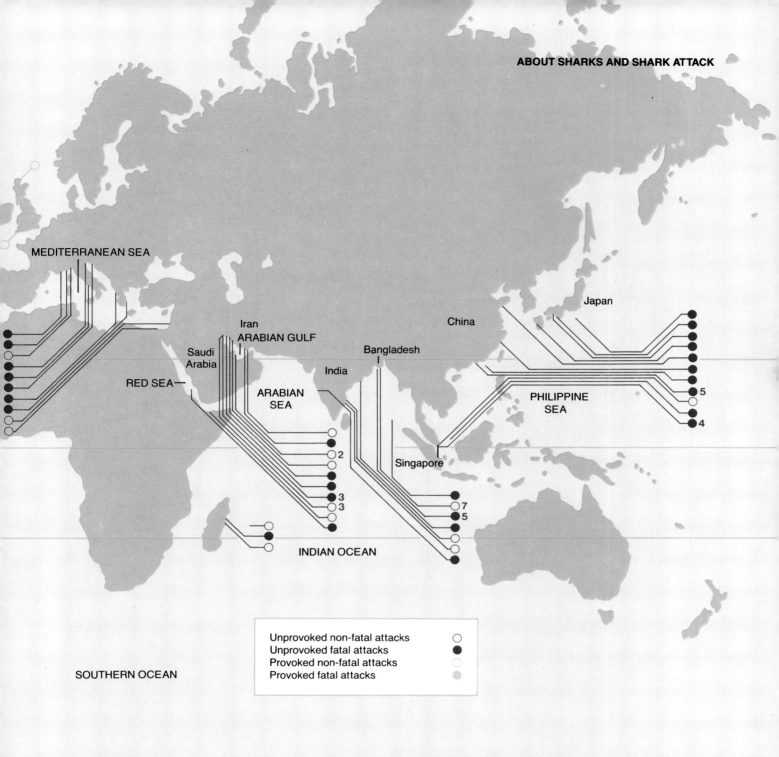

MEDITERRANEAN SEA

Japan

China

Iran
ARABIAN GULF

Saudi
Arabia

Bangladesh

RED SEA

India

ARABIAN
SEA

PHILIPPINE
SEA

5

4

2

3
3

Singapore

7
5

INDIAN OCEAN

SOUTHERN OCEAN

Unprovoked non-fatal attacks	○
Unprovoked fatal attacks	●
Provoked non-fatal attacks	○
Provoked fatal attacks	○

Attacks in the Caribbean Sea show little seasonal weighting – not surprising for a tropical region of the world. However, almost half the Caribbean attacks were fatal – a higher proportion than the world average.

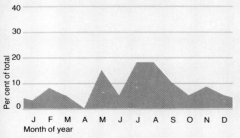

Per cent of total

J F M A M J J A S O N D
Month of year

The monthly breakdown of attacks in the Mediterranean Sea shows an expected bias towards the warmer months of the year – 90 per cent took place in July, August and September, when many people are swimming.

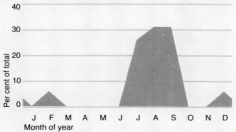

Per cent of total

J F M A M J J A S O N D
Month of year

GENERAL INDEX

Page numbers in bold type indicate main entries, while those in italic refer to illustrations. See separate species index on pages 202 to 207 for names that appear in the species charts (pages 156 to 179).

AB

Aboriginal paintings of sharks *64, 65*
Abyssal zone 13
Adriatic Sea 99, 196
African shark attacks 192-93
Ageing of sharks **47**
Agnatha 16
Ahvāz 196
Ainslie, W. 150
Air bladder, in fishes 16
Aitutaki 127
Albany 182
Aldinga Bay 84, 87
Almond, E.H. 100
Alopias pelagicus 20
Alopias superciliosus 20
Alopias vulpinus 20, *180*
American Indian shark myths 66
American Museum of Natural History 120
Ampullae of Lorenzini 22, *35*, **35-36**
Anal fin 23
Anatomy of fishes 16
Anatomy of sharks 17, 22
Ancient sharks **38-43**
Andromeda legend 60
Angel shark 19, 42, 68
Anglerius, Peter Martyn 66
Arandaspis 38
Aristotle 62-63, *63*, 184
Asai, Iona *87*
Asbury Park 120
Attacks on humans - see Shark Attacks
Attacks on boats by sharks 102
Auditory capsule of sharks 22
Auki 111
Australian Aborigines and sharks 64-65
Australian shark attacks **188-89**
Bakers Beach 95, 195
Baldridge, Dr David H. 72, 74, 78, 186
Ballentyne, R.M. 194
Banded sea snake 82, *83*
Bang stick *138*
Barlow, Roger 66
Barrett, John 123
Bartle, Bob 96-97
Basking shark 25, *145*, 148, 180
Basque Island 194
Bass Strait 188
Bathyl zone 13
Beach Haven 120
Beaurepaire, Frank 123
Bellerophon 69
Bellona Island 110
Benan Island 108
Benchley, Peter 71
Benia 106-107
Benthic sharks 20
Birkenhead 104, 105, **104-105,** 193
Birth of a shark *30*
Blackie, the shark catching dog *29*
Blacktip reef shark *17*
Blacktip shark 29
Bloch, Marcus Eliezer **67**, 70
Blood, human and sharks 81
Blue shark *18*, 20, 45, *46*, 47, 64, 71, *139, 143*, 184
Bonaparte, Napoleon *69*
Bondoc Peninsula 191
Bony fishes *16*

C

Bottlenose dolphin and sharks 29
Bource, Henri 50, 73, *90-91*
Braddock, Donald 182
Brady, Patrick 116-19, *119*
Brain of shark *32*
Brain to body-weight ratio of sharks *21*
Briggs, Captain Thomas 112
Brisbane River 189
Broken Bay 116
Brown shark 25, 34
Bruder, Charles 120
Bubble-curtain shark barriers 137
Buenos Aires 196
Bull shark 20, 24, 25, 27, *80*, 81, 125, 181, 183, 184
Bullhead sharks 19, 24
Buzzards Bay 94
Caldwell, Norman 147
Canadian shark attacks **194-95**
Canton Island 29
Cape Town 192
Caracalinga Head 95
Carcharhiniformes 18, 155, 167
Carcharhinus albimarginatus 82
Carcharhinus amblyrhynchos 24, *26*, 77, 82-83
Carcharhinus brachyurus 44
Carcharhinus galapagensis 47
Carcharhinus leucas 20, 27, *80*, 81, 125, 183, 184
Carcharhinus longimanus 68
Carcharhinus melanopterus *17*, 29
Carcharhinus plumbeus 25, 34, 47, 184
Carcharodon carcharias 20, 27, *80*, **50-53**, 146, 181, 183, 184, 188
Carcharodon megalodon 27, 42-43, *43*
Caribbean Sea 197
Carpet sharks 18, 24
Castaways and shark attacks **100-103**
Catsharks 24
Caudal fin 23
Caudal keel 23
Ceduna 182
Centrophorus squamosus 19
Cetorhinus maximus 148, 180
Challenger **70**
Chalmers, Jack 123
Charleston 92, 195
Chemical shark repellent *140*
Chimaeras 18
Chlamydosalachus anguineus 19, 42, 43, *70*
Chondrichthyes 18, 154
Cladoselache 40, 41
Claspers 26, *28*, 42, *71*
Co-operative shark tagging program 49
Coats of arms showing sharks *135*
Coledale 88
Columbus, Christopher 66
Compagno, Dr Leonard J.V. 154
Complete Natural History of Fish 67
Continental rise 13
Continental shelf 13
Continental slope 13
Coogee 114-17
Cook Islands 127
Cookiecutter shark 25, **27**, 28, 184
Copley, John Singleton 132-35
Copper shark *44*

DEF

Coppleson, Dr V.M. 78-79, 117
Copulation between sharks 28
Coquille 68
Corner, Jeff 95
Corwin, J.T. 37
Cottrell, Captain Thomas 121-22
Cow shark 19, 42
Cronulla *118*
Curaçoa 112
Cyclostomes 16
Dagon 109
Danger Point 104, 193
Dangerous Reef 52
Darwin, Charles 71
Davis, Dr D.H. 78
Dean, Alf 50, *180*, 182
Deep water sharks 20
Degei 108
Dequwaqa 108
Denticles *17*
Dijkgraaf, Sven 35-36
Doak, Wade 111
Doctor in Paradise, A 109
Dog-fish 64
Dog-shark 65
Dogfish *65*, 70, 135
Dogfish sharks 19
Dolphins and sharks 29
Dorsal fin 23
Dunn, Joseph 123
Duperry, Louis Isidore 68
Durban 92, 193
Dusky shark 32
Dusky smoothhound 37
Dwarf shark 20, *56*
Dyer, Bob *149, 185*
Dyer, Dolly *149*
Ears of sharks **37**
Eggcases of sharks *28*
Elasmobranchii 18, 43
Electrical receptors in sharks 28, **35-36**
Electrical shark barriers 137
Electrical shark repellers 138
Elephant fish 18
Ellis, William 65
Enclosures, shark proof *137*
Endolymphatic duct *37*
Eridacnis radcliffei 180
Eugomphodus taurus 24, *33*
Eusphyra blochii 183
Evolution of modern sharks *41*
Evolution of sharks and fishes *17*
Evolutionary history of sharks **38-43**
Exploitation of sharks **144-49**
Eyes of sharks 22, **34-35**
Facts about sharks **180-85**
Fake shark attacks 95
Farley, Bruce 90
Fastest sustained trip by a shark 45-46
Fay, R.R. 37
Ferret, HMS 112
Fijian legends about sharks 108-109
Fin spines 41
Fin structures of sharks 23
Firth of Forth 182
Fisch-Buch 10, 58, 152
Fish scales *16*
Fisher, Stanley 122-25
Fitton, Lieutenant Michael 112

GH

Flake 146
Food chains *15*
Food for sharks 26
Fossil sharks **38-43**
Fossilised shark teeth *39, 41, 43, 70*
Fox, Rodney 50, *86*, 87
Fraser, John 113
Fresh water sharks 184
Frilled shark *19, 42, 43, 70*
Galapagos shark *14, 47*, 80
Galeocerdo cuvier 24, *80*, 81, *126*, 182, 183, 184
Galeorhinus galeus 144, 146, 184
Galluchat, Jean-Claude 148
Game fishing records for sharks 185
Gamefish tagging programme 49
Ganges River 68, 184, 196
Ganges River shark 20, 184
Gas injection dart 138
Gases in seawater 13
Geological time scale *38*
Georges River 189
Gesner, Conrad 10, 58, 152
Gestation period of sharks 26
Gibbon, Wally *129*
Gilbert, Dr Perry W. 72, 137
Gill arches 22
Gill slits of sharks 17, 22
Gills 16, 17
Ginglymostoma cirratum 20, 34
Girardot, Lieutenant Frank 105
Glossopetrae 39, 150
Glyphis gangeticus 184
Gnathostomata 16
Goblin shark *19, 42, 182*, 183
Goldsmith, Oliver 70
Great Detached Reef 82
Great white shark 20, *25, 27, 33*, 48, **50-53**, *60, 67*, 75, *80*, 81, 85, *89, 93*, 125, 146, *151, 180, 181*, 183, 184, *185*, 188
Green, Clive 93
Greenland shark 147, 182
Grey nurse shark *8*, 24, 26, *33, 84*
Grey reef shark 24, *26*, 37, *77, 83*
Grey, Zane *148*
Ground sharks 18
Growth of sharks **47**
Guadalcanal 110
Gulf of St Lawrence 194
Gulf Stream 100
Gurr, James 93
Hadal zone 13
Hagfishes 16
Haiti 112
Hàkall 147
Halieutica 64
Hallstrom, Sir Edward 117
Hamelin Bay 188
Hammerhead shark *9*, 32, 66, *136*, 183
Harding, John 55
Hass, Hans 130
Hathaway, Marcia *98*
Havana harbour 132
Hawaii 66, 191
Hawkins, Sir John 68, 103
Haye 68
Head, Detective Constable Frank 114
Heart of shark 23
Helicoprion 42

IJKL

Hemingway, Ernest 71
Herbert, Sir Thomas 67
Herodotus 60
Heron Island 128, *129*
Heterodontiformes 19, 155, 162
Heteroduntus portusjacksoni 9, 19, 26, 28, 43, 54, 183
Hexanchidae 43
Hexanchiformes 19, 155, 156
Historia Animalium 62, 184
History of sharks and humans **60-71**
Hobson, Charles and Albert *115*
Holmes, Harry 97
Holmes, Reginald 116-19, *119*
Holocephali 18, 43
Homer, Winslow 101
Hong Kong 196
Hooghly River 184, 196
Hopper, Jack 98
Horley, Philip *85*
How to identify sharks 155
Hybodus 40, 42
Hydobonts 42
Identification chart of sharks 155
Iniopterygians 43
Institute of Jamaica 113
International Game Fishing Association 150
Intestines of sharks 17
Invertebrates 14
Inyoni Rocks 93
Isistius brasiliensis 27, 27-28, 184
Isurus oxyrinchus 20, 22-23, 184, 227
Isurus paucus 20
Jamaica 112
Jaws 71, 93
Jewellery from shark teeth *146*
Jonah and the whale (shark) *60-61*
Jurien Bay 96, 96-97
Kabat, Lieutenant Commander 102
Kadavu 108
Kalmijn, A.J. 35-36
Kapata 65
Karun River 196
Keel harbour *144*
Keeping sharks at bay **136-43**
Kelly, Vince
Kendrick, Damon 92
Kevlar anti-shark suit *139*
Kiribati 29, 107
Kiriwina 106
Kogler, Albert 95, 195
La Jolla Cove 93
Lake Nicaragua shark 20
Lambert, Dr S.M. 109
Lamna ditropis 20
Lamna nasus 20, 146, 184
Lamniformes 19, 155, 165
Lampreys 16
Langalanga Lagoon 108
Langenorhynchus obliquidens 25
Lateral cutaneous artery 23
Lateralis system in sharks *36*, 36-37
Lau Lagoon 108
Laulasi 110
Leafscale gulper shark *19*
Legauat, Francois 69
Lehrer, Gerald 93

MN

Lemon shark 27, *30*, 32, 34, 138, 181
Leopard shark 54, *78*
Lerner Marine Laboratory 32, 34, 137
Levuka 108
Life magazine 95
Life span of sharks 183
Lineaweaver, Thomas A. 53
Linnaeus, Carl 154
Littoral zone 13
Liver of sharks 17, 23, 25, 144, 147-48
Lives and habits of sharks **20-31**
Longnose sawshark *18*
Los Angeles 195
Lucas, Dr Frederick A. 120
Macbeth 68
Mackay 182
Mackerel sharks 19
Macula neglecta 37
Magellan, Ferdinand 66
Magnus, Olaus 67
Mako shark 20, *22-23*, 27, 45, 46, *149*, 151, 184
Malaita 108, 110
Mano-kanaka 65
Manrique, Sebastian 68-69
Maoris and sharks 65
Marine Biological Laboratory 32
Marine Industries Ltd 147
Matawan **120-25**, 195
Materia Indica 150
Mating of sharks 26, 28
Mating scars on sharks *28*
Maxwell, Gavin 148
McEwan, Norman 191
Mediterranean Sea 196-97
Megachasma pelagios 20, 28, *183*
Megamouth shark *20*, 28, *183*
Melville, Herman 71
Mental abilities of sharks 32
Mesh anti-shark suit 139, **142-43**
Meshing 136-37, *137*
Mexican shark attacks **194-95**
Miami 195
Mid-water sharks 20
Middle Harbour 189
Migration of sharks 21, **45-46**
Mitsukurina owstoni 19, 183
Moby Dick 71
Monk-fish 68
Moreton Bay 189
Moses sole *141*
Most travelled species of shark 46
Mote Marine Laboratory 181, 186
Mundus, Frank 182
Murray, R.W. 35
Muscle structure of sharks 23
Nancy 112
Natal 93, 193
National Marine Fisheries Service 45
Negaprion brevirostris 27, 32, 139
Negros 191
Nekton 15
Neoselachians 41, 43
New Georgia Sound 110
New Jersey shark attacks (1916) **120-25**
New Zealand shark attacks **190-91**
Nichols, Dr A.T. 120
Noah 66
Nostrils of sharks 22, **32-33**, *33*

Nova Scotia 103, 193
Novak, Agnes 196
Nurse shark 20, *21*, 24, 25, *28*, 34, 37, 42
Ocean Leather Corporation 144
Oceans, properties of 12-14
Odontaspididae 42
Oil from shark livers **25**, 144, 147-48
Old Man and the Sea, The 71
Oppianus 64
Orbeck, Peter 70, 144
Orectolobiformes 18, 155, 163
Orectolobus ornatus 18
Ornate wobbegong *18*
Osprey Reef 82
Osteichthyes 18
Ostle, Colin *181*
Otoliths 37
Pacific Ocean *12*
Pacific Ocean shark attacks **190-91**
Pacific sleeper shark 182
Pacific whiteside dolphin 25
Palmer, Dr Arthur 115
Pamperin, Robert L. 93
Parasites of sharks 24
Pardachirus marmoratus 141
Parker, George H. 36
Pearl divers and sharks 64-65
Pectoral fin 23
Pelagic sharks 20
Pelvic fin 23
Persian Gulf 196
Phillips, Allen 85
Phytoplankton 15
Pikaea 38
Piked dogfish 146, 182, 184
Pilot fish 183
Pindimar 147
Pit organs of sharks 37
Placoderms 40
Plagiostomi 155
Plankton 15
Pliny the Elder 64
Polo, Marco 65, 66
Pomet, Monsieur 69, 150
Pompeii 62
Pope, Ratu 109
Popper, A.N. 37
Porbeagle shark *25*, 42, 146, 184
Port Arthur 188
Port Davey 188
Port Fairey 182
Port Jackson shark *9*, *19*, 26, 27, *28*, 43, 54, 183
Portsea 98
Port Royal 112
Powerhead *138*
Prionace glauca 18, 20, 45, *46*, *143*, 184
Pristiophoriformes 18, 155, 161
Pristiophorus cirratus 18
Protractor lentis 35
Pteraspis 38
Purchas, Samuel 68
Pygmy ribbontail catshark 180
Ra's At-tannūrah 196
Rabbit-fishes 18
Raritan Bay 125
Rat-fishes 18
Reading, Lieutenant A.G. 100
Records, game fishing 185

Remora 30, *183*
Requiem 68
Rhiniodon typus 20, **54-57**, 180
Rhodes, Cecil 71
Risk of shark attack **72-81**
Rodger, Brian 50, 84
Rolliad, The 135
Rondelet, Guillaume 60, 68
Rose, K.J. 35
Roviana Lagoon 108
Ruhen, Olaf 106
Sacculus 37
Sailors and sharks *103*
Salmon shark 20
Salmond, Captain 104
San Diego 195
San Francisco 195
Sand devil *19*
Sand shark 42
Sandbar shark 46, 47, 184
Sandtiger (grey nurse) shark *8*, 24, 26, *33*, *84*
Sandy Point 97
Savo Island 108
Sawshark 18, 42
Scales, fish *16*
Scarr, Deryck 109
Schapper, Hans 196
School shark *151*
Schultz, Dr Leonard 72
Scyliorhinus canicula 35
Sea snake, banded *83*
Sea snakes and sharks 82-83, *83*, 138
Sea-cow *65*
Sea-dog *65*, 68
Sea-horse *65*
Seals and sharks 73
Seawater, chemical composition of 12
Selache 62
Selachimorpha 155
Sense of smell in sharks 53
Senses of sharks **32-37**
Seton, Lieutenant Colonel Alexander 104
Sevengill sharks 43
Shagreen 148
Shakespeare, William 68
Shark Arm Mystery, The **114-19**
Shark Attack 79
Shark Attack File 74, 186
Shark attacks
   accounts of **84-105**
   and divers 74,77
   and human blood 81
   and seals 73
   Birkenhead **104-105**
   Brook Watson **132-35**
   characteristics of 79
   charmed lives of rescuers 123
   country of attack *78*
   day of week of attack *78*
   distance from shore of incidents 74, 76
   effects of race on 76
   effects of water conditions on 74
   effects of water temperature on 78
   factors that determine the risk of 74-75
   faked *95*

first aid measures 81
first documented case 68
global patterns of **186-97**
in Africa **192-93**
in Asia **196-97**
in Australia **188-89**
in Canada **194-95**
in Europe **196-97**
in New Zealand **190-91**
in South Africa **192-93**
in South America **196-97**
in the Caribbean Sea **196-97**
in the Mediterranean Sea **196-97**
in the Pacific Ocean **190-91**
in the United States **194-95**
influence of colour of clothing on 76
influence of water depth on 74
most dangerous species 80
most northerly 196
most southerly 191
New Jersey (1916) **120-25**
on boats *102*
on castaways 100-105
on pearl divers 87
on surfboard riders 73, *85*
pattern of **186-97**
ratio of males to females 73
reducing the risk of 81
risk of **72-81**
time of day of *78*
types of injuries 80, *86*, *90*
Shark Attacks Against Man 72
Shark Bay 129
Shark charmers 65, **106-111**
Shark charms *108*
Shark Chaser *141*
Shark corneas 150
Shark curios 150
Shark fin soup 137, **146**
Shark fishing contests *137*
Shark leather 144, 150
Shark Panel 72
Shark Papers, The **112-13**
Shark Screen *140*
Shark tooth swords 108
Shark-Man 66
Sharks
   ageing 47
   ampullae of Lorenzini 22, **35-36**
   anal fin 23
   anatomy of *17*, *22*
   ancient **38-43**
   and divers *87*
   and dolphins 29
   and sea snakes **82-83**
   and surfboard riders 73, **85**
   as food 66, 137, 144, 151
   assessments of stocks 47
   auditory capsule 22
   benthic 21
   bite force 28
   blood supply 23
   body language 77
   body temperature 53
   brain *32*
   brain to body-weight ratio *21*
   buoyancy 25
   cartilage 17
   caudal fin 23

caudal keel 23
charming 106-111
copulation 28
dangerous *80*, 183
deep water 21
denticles *23*
deterrents **136-41**
dorsal fin 23
dried fins *146*
ears *37*
eating habits 117
effects of temperature on
 behaviour 21
egg cases *28*
eggs 26
electrical receptors 28, **35-36**
evolutionary history **39-43**
exploitation **144-51**
eyes 22, **34-35**
facts **180-85**
fastest swimmer 184
fastest trip by 45
fin structure 23
fishing records 185
food 26
fossil **38-43**
fresh water species 184
gestation period 26
gill arches 22
gills 17, 22
giving birth *30*
greatest travellers 184
growth rates 47
growth rings in vertebrae *47*
heart 23
hides for leather *144*
identification chart 155
in fiction 71
in history **60-71**
in ocean depths 184
jewellery *146*
largest **54-57**, 180
largest maneater 180
lateral cutaneous artery 23
lateralis system **36-37**
learning ability 21
leather from 137
life span 183
list of species **154-79**
liver 17, 23, *25*
lives and habits **20-31**
mating 26-28
medical remedies from 69
mental abilities 32
methods of bearing young 26
methods of swimming 24
mid-water 21
migrations 21, *45*
most common 182
most dangerous 183
most travelled species 46
muscle structure 23
myths and legends **60-69**, 106-111
nostrils 22, **32-33**
oil in liver 25
on coats of arms 135
orders of 18-19
origin of name 68
pancreas 23

parasites 24
pectoral fin 23
pelagic 21
pelvic fin 23
pit organs 37
poisonous species 147
popular myths 184
power of 24
protection against attack from **136-43**
rarest 182
record movements 46
records **180-85**
risk of attack by **72-81**
segregation of sexes 47, 52
sense of smell 32, 53
senses **32-37**
sensitivity to electrical currents 35-36
sensitivity to magnetic fields 36
sensitivity to vibration 36
skeleton 22
skin 23
sleeping 24
smallest *56*, 180
species of **154-79**
species that attack humans 80, 183
spiracle 22
spiral valve 17, 23
stomach 23
strangest 183
supposed preference for human
 flesh 69
swimming speeds 24, 184
tagging programmes **44-49**
tail 17, *25*
teeth 17, *22*, *27*
tooth shapes *27*
warm blooded 21
worship of 106-111
Sharks of the World 154
Sharkskin 148
Sheldon, Ralph E. 32
Shipwrecks and sharks **100-105**
Short, Raymond *88*
Silvertip shark 82
Singapore 196
Sixgill sharks 43
Small-spotted catshark 35
Smith, James *116*, 116-19
Society Islands 65
Solomon Islands 65, 110
Somniosus microcephalus 147, 182
Somniosus pacificus 182
Soupfin shark 144, 146
South African shark attacks **192-93**
South American shark attacks **196-97**
Sparrow, HMS 112
Spear gun 138
Species of sharks **154-79**
Spence, Geoffrey 93
Spined pygmy shark 56, 180
Spiny dogfish 24
Spiny sharks 38
Spiracle of sharks 22
Spiral valve of sharks 47
Spot-tail shark 47
Spring Lake 120
Squalene 148
Squaliformes 19, 155, 156
Squaliolus laticaudus 20, *56*, 180

T-Z

Squalus acanthias 24, 146, 182, 184
Squatina dumeril 19
Squatiniformes 19, 155, 161
St Georges Channel 107
Stethacanthus 39, 40
Stiles, Walter 94
Stilwell, Lester 121-22
Stomach of sharks 23
Striped wet suit experiments **82-83**
Sublittoral zone 13
Sugarloaf Bay 98
Sukuna, Ratu Sir Lala 109
Sullivan, Jeremiah 139, 142-43
Surfboard riders and sharks *73*, *85*
Sydney Harbour 94, *98*, 189
Sydney shark attacks 188-89
Tagging programmes **44-49**
Tails of sharks *25*
Tapetum lucidum 34
Taveuni 108
Taylor, Ron and Valerie 8-9, 50-53, *55*,
 82-83, *85*, 126-31, 138, *142-43*
Teeth, fossilised *39*, *41*
Thelodonts 40
Thelodus 38
Thresher shark 20, 24, *25*, *180*
Tiburon 68, 73
Tiger shark 24, *25*, 26, *27*, 45, 73, 81,
 112, 114, *117*, 121, 125, *126*, 128-29,
 149, 182, 183, 184, *185*
Tonga 65, 107
Tope shark 146, 183
Torres Strait 87
Treacle (pearl diver) 87
Triaenodon obesus 83, 129, 130, 183
Triakis semifasciata 54
Trieste *99*
Trobriand Islands 106
Troy, Joseph 94
Tulaghi 110
Tursiops truncatus 29
Twain, Mark 71
United States shark attacks **194-95**
Vansant, Charles E. 120
Vaughan, Greg 116-19
Vertebrates 14
Vishnu 66
Vitamin A from shark liver oil 148
Warland, Gregory 98
Warner, Lee 96-97
Waterhouse, Reverend J. 108
Watson, Sir Brook **132-35**
Whale shark 20, 26, 29, **54-55**, 180
White pointer – see great white shark
Whitetip reef shark *79*, *83*, *129*, 130,
 183
Wick 196
Wilkes, John 135
Wilson, President Woodrow 125
Winklespruit Beach 92
Wishart, John 98
Wobbegong *9*, *18*
Wright, Captain 105
Wright, Richard 98
Wylie, Lieutenant Hugh 112
Xenacanthids 40
Zambezi River shark 20
Zooplankton *14*, 15

SPECIES INDEX

The following index lists both the common and scientific names of all the shark species that appear in the charts on pages 156 to 179. Alternative common names, and some outdated scientific names are also included.

AB

Aculeola nigra 156
African angelshark 161
African lanternshark 158
African ribbontail catshark 172
African sawtail catshark 169
African spotted catshark 170
alligator dogfish 156
Alopias pelagicus 166
Alopias superciliosus 166
Alopias vulpinus 166
alopiidae 166
American sawshark 161
angel fish 162
angel shark 162
angel sharks 161-2
angular roughshark 161
Aprionodon brevipinna see
 Carcharhinus brevipinna 176
Aprionodon isodon see *Carcharhinus isodon* 176
Apristurus atlanticus 167
Apristurus brunneus 167
Apristurus canutus 167
Apristurus herklotsi 167
Apristurus indicus 167
Apristurus investigatoris 167
Apristurus japonicus 167
Apristurus kampae 167
Apristurus laurussoni 167
Apristurus longicephalus 167
Apristurus macrorhynchus 167
Apristurus maderensis 167
Apristurus manis 167
Apristurus microps 167
Apristurus nasutus 168
Apristurus parvipinnis 168
Apristurus platyrhynchus 168
Apristurus profundorum 168
Apristurus riveri 168
Apristurus saldanha 168
Apristurus sibogae 168
Apristurus sinensis 168
Apristurus spongiceps 168
Apristurus stenseni 168
Apristurus verweyi 168
Arabian carpetshark 164
Arabian catshark 169
Arabian smooth-hound 174
Argentine angel shark 161
arrowhead dogfish 157
Asymbolus analis 168
Asymbolus vincenti 168
Atelomycterus macleayi 168
Atelomycterus marmoratus 168
Atlantic angel shark 161
Atlantic ghost catshark 167
Atlantic sharpnose shark 179
Atlantic weasel shark 175
Aulohalaelurus labiosus 168
Australian angel shark 161
Australian marbled catshark 168
Australian sawtail catshark 169
Australian sharpnose shark 179
Australian spotted catshark 168
Australian swellshark 168
Bahamas sawshark 161
balloon shark 169
bamboo sharks 164
banded catshark 170

banded houndshark 175
banded wobbegong 163
barbeled catshark 171
barbeled houndshark 172
barbeled houndsharks 172
barbelthroat carpetshark 163
bareskin dogfish 157
basking shark 166
basking sharks 166
bigeye houndshark 173
bigeye sand tiger 165
bigeye thresher 166
bigeyed sixgill shark 156
bignose shark 175
birdbeak dogfish 157
black dogfish 157
black shark (for combtooth dogfish, also
 kitefin shark) 157
black whaler 177
blackbelly dogfish 158
blackbelly lanternshark 158
blackgill catshark 170
blackmouth catshark 169
blackmouthed dogfish 169
blacknose shark 175
blackspot shark 177
blackspotted catshark 168, 170, 171
blackspotted smooth-hound 174
blacktail reef shark 177
blacktailed spurdog 160
blacktip reef shark 177
blacktip sawtail catshark 169
blacktip shark 177
blacktip tope 173
Blainville's dogfish 160
blind shark 163
blind sharks 163
blotched catshark 172
blue pointers 166
blue shark 178
blue whaler 178
blue-dog 178
bluegrey carpetshark 163
bluespotted bambooshark 164
bluespotted catshark 164
bluntnose sixgill shark 156
boa catshark 171
bone shark 166
bonito shark 166
Bonnaterre's deepwater shark 157
bonnethead 179
bonnethead sharks 179
Borneo catshark 168
Borneo shark 176
brachaeluridae 163
Brachaelurus waddi 163
bramble shark 156
bramble sharks 156
Brazilian shark 159
Brazilian sharpnose shark 178
bristly catshark 170
broadbanded lanternshark 158
broadfin sawtail catshark 169
broadfin shark 178
broadgill catshark 168
broadnose catshark 167
broadnose sevengill shark 156
broadsnout 156
brown catshark 163, 167, 170

C

brown lanternshark 158
brown shyshark 170
brown smooth-hound 174
brownbanded bambooshark 164
brownspotted catshark 164, 171
bull huss 172
bull shark 176
bulldog shark 156, 162
bullhead sharks 162
calf shark 156
Campeche catshark 170
carcharhinidae 175-9
Carcharhiniformes 167-79
Carcharhinus acronotus 175
Carcharhinus ahenea see *C. brachyurus* 176
Carcharhinus albimarginatus 175
Carcharhinus altimus 175
Carcharhinus amblyrhynchoides 175
Carcharhinus amblyrhynchos 175
Carcharhinus amblyrhynchus see *C. wheeleri* 177
Carcharhinus amboinensis 176
Carcharhinus azureus see *C. leucas* 176
Carcharhinus bleekeri see *C. plumbeus* 177
Carcharhinus borneensis 176
Carcharhinus brachyurus 176
Carcharhinus brevipinna 176
Carcharhinus cautus 176
Carcharhinus dussumieri 176
Carcharhinus falciformis 176
Carcharhinus fitzroyensis 176
Carcharhinus floridanus see *C. falciformis* 176
Carcharhinus galapagensis 176
Carcharhinus gangeticus see *Glyphis gangeticus* 178
Carcharhinus glyphis see *Glyphis glyphis* 178
Carcharhinus hemiodon 176
Carcharhinus isodon 176
Carcharhinus japonicus see *C. plumbeus* 177
Carcharhinus lamiella see *C. obscurus* 177
Carcharhinus leucas 176
Carcharhinus limbatus 177
Carcharhinus longimanus 177
Carcharhinus macloti 177
Carcharhinus macrurus see *C. obscurus* 177
Carcharhinus maculipinnus see *C. brevipinna* 176
Carcharhinus maou see *C. longimanus* 177
Carcharhinus melanopterus 177
Carcharhinus menisorrah see *C. amblyrhynchos* 175 or *C. dussumieri* 176
Carcharhinus milberti see *C. plumbeus* 177
Carcharhinus nicaraguensis see *C. leucas* 176
Carcharhinus obscurus 177
Carcharhinus oxyrhynchus see *Isogomphodon oxyrhynchus* 178
Carcharhinus perezi 177
Carcharhinus platyodon see *C.*

plumbeus 177
Carcharhinus platyrhynchus see *C. albimarginatus* 175
Carcharhinus pleurotaenia see *C. amblyrhynchoides* 175
Carcharhinus plumbeus 177
Carcharhinus porosus 177
Carcharhinus remotus see *C. brachyurus* 176
Carcharhinus sealei 177
Carcharhinus signatus 177
Carcharhinus sorrah 177
Carcharhinus spallanzani see *C. wheeleri* 177
Carcharhinus springeri see *C. perezi* 177
Carcharhinus stevensi see *C. plumbeus* 177
Carcharhinus temmincki see *Lamiopsis temmincki* 178
Carcharhinus vanrooyeni see *C. leucas* 176
Carcharhinus wheeleri 177
Carcharhinus zambezensis see *C. leucas* 176
Carcharhias arenarius see *Eugomphodus taurus* 165
Carcharhias ferox see *Odontaspis ferox* 165
Carcharias platensis see *Eugomphodus taurus* 165
Carcharias sorrakowah see *Scoliodon laticaudus* 179
Carcharias taurus see *Eugomphodus taurus* 165
Carcharodon carcharias 166
Caribbean lanternshark 158
Caribbean reef shark 177
Caribbean roughshark 161
Caribbean sharpnose shark 178
carpetshark 163
carpetsharks 163-5
catsharks 163, 167-72
Centrophorus acus 156
Centrophorus granulosus 156
Centrophorus harrisoni 156
Centrophorus lusitanicus 156
Centrophorus moluccensis 156
Centrophorus niaukang 156
Centrophorus scalpratus see *C. moluccensis* 156
Centrophorus squamosus 156
Centrophorus tessellatus 157
Centrophorus uyato 157
Centroscyllium fabricii 157
Centroscyllium granulatum 157
Centroscyllium granulosum see *C. nigrum* 157
Centroscyllium kamoharai 157
Centroscyllium nigrum 157
Centroscyllium ornatum 157
Centroscyllium ritteri 157
Centroscymnus coelolepis 157
Centroscymnus crepidater 157
Centroscymnus cryptacanthus 157
Centroscymnus macracanthus 157
Centroscymnus owstoni 157
Centroscymnus plunketi 157
Cephaloscyllium fasciatum 168
Cephaloscyllium isabellum 168

D

Cephaloscyllium laticeps 168
Cephaloscyllium nascione 169
Cephaloscyllium silasi 169
Cephaloscyllium sufflans 169
Cephaloscyllium ventriosum 169
Cephalurus cephalus 169
cetorhinidae 166
Cetorhinus maximus 166
Cetorhinus normani see *C. maximus* 166
Cetorhinus rostratus see *C. maximus* 166
Chaenogaleus macrostoma 175
chain catshark 172
Chiloscyllium arabicum 164
Chiloscyllium caerulopunctatum 164
Chiloscyllium colax see *C. indicus* 164
Chiloscyllium griseum 164
Chiloscyllium indicum 164
Chiloscyllium plagiosum 164
Chiloscyllium punctatum 164
chlamydoselachidae 156
Chlamydoselachus anguineus 156
cigar shark 159
Cirrhigaleus barbifer 157
Cirrhoscyllium expolitum 163
Cirrhoscyllium formosanum 163
Cirrhoscyllium japonicum 163
clouded angel shark 162
cloudy catshark 172
cobbler carpetshark 164
cobbler wobbegong 164
cocktail shark 176
codshark 160
Colclough's shark 163
collared carpetshark 163
collared carpetsharks 163
combtooth dogfish 157
combtoothed lanternshark 158
common sawshark 161
Cooke's shark 156
cookiecutter shark 159
copper shark 176
coral catshark 168
cow sharks 156
creek whaler 176
crested bullhead shark 162
crested Port Jackson shark 162
crocodile shark 166
crocodile sharks 166
Ctenacis fehlmanni 172
cub shark 176
Cuban dogfish 160
Cuban ribbontail catshark 172
Cyrano spurdog 160
daggernose shark 178
Dalatias licha 157
Dalatias phillipsi see *D. licha* 157
dark shyshark 170
Deania calcea 157
Deania histricosa 157
Deania profundorum 157
Deania quadrispinosum 158
death shark 166
deepwater catshark 168
dogfish sharks 156-61
draughtsboard shark 168
draughtsboard sharks 168
dumb gulper shark 156
dumb shark 172

EF

dusky catshark 170
dusky shark 177
dusky smooth-hound 173
dwarf catshark 172
dwarf sawtail catshark 169
echinorhinidae 156
Echinorhinus brucus 156
Echinorhinus cookei 156
elephant shark 166
elfin shark 165
Endeavour dogfish 156
epaulette shark 164
epaulette sharks 164
Eridacnis barbouri 172
Eridacnis radcliffei 172
Eridacnis sinuans 172
Etmopterus baxteri 158
Etmopterus brachyurus 158
Etmopterus bullisi 158
Etmopterus decacuspidatus 158
Etmopterus gracilispinis 158
Etmopterus granulosus 158
Etmopterus hillianus 158
Etmopterus lucifer 158
Etmopterus polli 158
Etmopterus princeps 158
Etmopterus pusillus 158
Etmopterus schultzi 158
Etmopterus sentosus 158
Etmopterus spinax 158
Etmopterus unicolor 158
Etmopterus villosus 158
Etmopterus virens 158
Eucrossorhinus dasypogon 163
Eucrossorhinus ogilbyi see *E. dasypogon* 163
Eugomphodus taurus 165
Eugomphodus tricuspidatus 165
Eulamia gangetica see *Glyphis gangeticus* 178
Eulamia macrura see *Carcharhinus obscurus* 177
Eulamia milberti see *Carcharhinus plumbeus* 177
Eulamia temmincki see *Lamiopsis temmincki* 178
Euprotomicroides zantedeschia 158
Euprotomicrus bispinatus 159
Euprotomicrus laticaudus see *Squaliolus laticaudus* 160
Eusphyra blochii 179
false catshark 172
false catsharks 172
Figaro boardmani see *Galeus boardmani* 169
filetail catshark 171
finback catsharks 172
finetooth shark 176
flake 173
flapnose houndshark 174
flapnosed smooth-hound 174
flathead catshark 167
Florida dogfish 174
fox shark 166
freckled catshark 172
Freycinet's shark 164
frill shark 156
frilled shark 156
frilled sharks 156

G

frilled-gilled shark 156
fringefin lanternshark 158
Furgaleus macki 173
Furgaleus ventralis see *F. macki* 173
Galapagos bullhead shark 162
Galapagos shark 176
Galeocerdo arcticus see *G. cuvier* 177
Galeocerdo cuvier 177
Galeolamna macrurus see *Carcharhinus obscurus* 177
Galeolamna stevensi see *Carcharhinus plumbeus* 177
Galeorhinus australis see *G. galeus* 173
Galeorhinus chilensis see *G. galeus* 173
Galeorhinus galeus 173
Galeorhinus omanensis see *Iago omanensis* 173
Galeorhinus zyopterus see *G. galeus* 173
Galeus arae 169
Galeus boardmani 169
Galeus eastmani 169
Galeus melastomus 169
Galeus melastomus murinus see *G. murinus* 169
Galeus murinus 169
Galeus nipponensis 169
Galeus piperatus 169
Galeus polli 169
Galeus sauteri 169
Galeus schultzi 169
Ganges shark 176, 178
gecko catshark 169
ghost catshark 167
giant sleepy shark 165
Ginglymostoma brevicaudatum 165
Ginglymostoma cirratum 165
Ginglymostoma ferrugineum see *Nebrius ferrugineus* 165
ginglymostomatidae 165
Glyphis gangeticus 178
Glyphis glyphis 178
goblin shark 165
Gogolia filewoodi 173
golden dogfish 157
Gollum attenuatus 172
graceful catshark 172
graceful shark 175
granular dogfish 157
great blue shark 178
great hammerhead 179
great lanternshark 158
great white shark 166
green dogfish 158
green lanternshark 158
Greenland shark 159
grey bambooshark 164
grey nurse shark 165
grey reef shark 175
grey reef whaler 176
grey shark 156
grey sharpnose shark 178
grey smooth-hound 173
ground sharks 167-79
Gulf catshark 168
gully shark 175
gulper shark 156
gulper sharks 156
gummy shark 173
gurry shark 159

H

Halaelurus alcocki 169
Halaelurus analis see *Asymbolus analis* 168
Halaelurus bivius see *Schroederichthys bivius* 171
Halaelurus boesemani 170
Halaelurus buergeri 170
Halaelurus canescens 170
Halaelurus chilensis see *Schroederichthys chilensis* 171
Halaelurus dawsoni 170
Halaelurus hispidus 170
Halaelurus immaculatus 170
Halaelurus lineatus 170
Halaelurus lutarius 170
Halaelurus natalensis 170
Halaelurus quagga 170
Halaelurus vincenti see *Asymbolus vincenti* 168
Halaelurus labiosus see *Aulohalaelurus labiosus* 168
Halsydrus maccoyi see *Cetorhinus maximus* 166
Halsydrus maximus see *Cetorhinus maximus* 166
hammerhead sharks 179
Haploblepharus edwardsii 170
Haploblepharus fuscus 170
Haploblepharus pictus 170
hardnose shark 177
hardnose smooth-hound 174
harlequin catshark 172
Hawaiian lanternshark 158
head shark 169
hemigaleidae 175
Hemigaleus microstoma 175
Hemipristis elongatus 175
hemiscylliidae 164
Hemiscyllium freycineti 164
Hemiscyllium hallstromi 164
Hemiscyllium ocellatum 164
Hemiscyllium strahani 164
Hemiscyllium tricuspidare see *H. trispeculare* 164
Hemiscyllium trispeculare 164
Hemitriakis japanica 173
Hemitriakis leucoperiptera 173
Heptranchias perlo 156
heterodontidae 162
Heterodontiformes 162
Heterodontus francisci 162
Heterodontus galeatus 162
Heterodontus japonicus 162
Heterodontus mexicanus 162
Heterodontus portusjacksoni 162
Heterodontus quoyi 162
Heterodontus ramalheira 162
Heterodontus zebra 162
Heteroscyllium colcloughi 163
Heteroscymnoides marleyi 159
Heteroscymnus longus see *Somniosus rostratus* 160
hexanchidae 156
Hexanchiformes 156
Hexanchus griseus 156
Hexanchus vitulus 156
hoary catshark 167
Holohalaelurus punctatus 170
Holohalaelurus regani 170

IJKL

hooded carpetshark 164
hooktooth dogfish 156
hooktooth shark 175
horn shark 162
horn sharks 162
horned sharks 162
houndsharks 173-5
humantin 160
humpback smooth-hound 174
huss, bull 172
Hypogaleus hyugaensis 173
Hypoprion macloti see *Carcharhinus macloti* 177
Hypoprion playfairi see *Carcharhinus melanopterus* 177
Hypoprion signatus see *Carcharhinus signatus* 177
Iago garricki 173
Iago omanensis 173
Iceland catshark 167
Indian sand tiger 165
Indian swellshark 169
Indonesian speckled carpetshark 164
Isabel's swellshark 168
Isistius brasiliensis 159
Isistius plutodus 159
Isogomphodon oxyrhynchus 178
Isurus glaucus see *I. oxyrinchus* 166
Isurus oxyrinchus 166
Isurus paucus 167
Izak 170
Izak catshark 170
Japanese angel shark 162
Japanese bullhead shark 162
Japanese catshark 167
Japanese sawshark 161
Japanese spurdog 160
Japanese topeshark 173
Japanese wobbegong 163
Java shark 176
Juncrus vincenti see *Asymbolus vincenti* 168
kitefin shark 157
knifetooth dogfish 159
Knopp's shark 176
Lamiopsis temmincki 178
Lamna ditropis 167
Lamna nasus 167
Lamna philippi see *L. nasus* 167
Lamna whitleyi see *L. nasus* 167
lamnidae 166-7
Lamniformes 165-7
lanternsharks 158
largenose catshark 168
largespine velvet dogfish 157
large-spotted dogfish 172
largetooth cookiecutter shark 159
leafscale gulper shark 156
lemon shark 178
lemon sharks 178
leopard catshark 171
leopard shark 164, 175
Leptocharias smithii 172
leptochariidae 172
lesser sixgill shark 156
lesser soupfin shark 173
lined catshark 170
lined lanternshark 158
little gulper shark 157

M

little sawshark 161
little sleeper shark 160
lollipop catshark 169
longfin catshark 167
longfin mako 167
longhead catshark 167
longnose catshark 167
longnose grey shark 176
longnose houndshark 173
longnose pygmy shark 159
longnose sawshark 161
longnose spurdog 160
longnose velvet dogfish 157
long-nosed black-tail shark 175
longsnout dogfish 158
longtailed carpetsharks 164
Lord Plunket's shark 157
lowfin gulper shark 156
Loxodon macrorhinus 178
luminous shark 159
mackerel shark 167
mackerel sharks 165-7
Madame X 165
Madeira catshark 167
makos 166-7
mandarin dogfish 157
maneater 166
Mapolamia spallanzanii see
   Carcharhinus melanopterus 177
marbled catshark 169
Megachasma pelagios 166
megachasmidae 166
megamouth shark 166
Mexican hornshark 162
milk shark 178
Mitsukurina owstoni 165
mitsukurinidae 165
Moller's deepsea dogfish 158
Molochophrys galeatus see *Heterodontus*
   galeatus 162
monkey-mouthed shark 164
monkfishes 161-2
monk sharks 161-2
mosaic gulper shark 157
Moses smooth-hound 174
mouse catshark 169
Mozambique bullhead shark 162
mud catshark 170
mud shark 156
Mustelus antarcticus 173
Mustelus asterias 173
Mustelus californicus 173
Mustelus canis 173
Mustelus dorsalis 173
Mustelus fasciatus 173
Mustelus griseus 173
Mustelus henlei 174
Mustelus higmani 174
Mustelus kanekonis see *M. griseus* 173
Mustelus lenticulatus 174
Mustelus lunulatus 174
Mustelus manazo 174
Mustelus mento 174
Mustelus mosis 174
Mustelus mustelus 174
Mustelus norrisi 174
Mustelus palumbes 174
Mustelus punctulatus 174
Mustelus schmitti 174

NOP

Mustelus whitneyi 174
narrowfin smooth-hound 174
narrowmouthed catshark 171
narrownose smooth-hound 174
narrowtail catshark 171
Nasolamia velox 178
Nebrius concolor see *N. ferrugineus* 165
Nebrius ferrugineus 165
necklace carpetshark 163
needle dogfish 156
Negaprion acutidens 178
Negaprion brevirostris 178
Negaprion fronto see *N. brevirostris* 178
Negogaleus microstoma see *Hemigaleus*
   microstoma 175
nervous shark 176
New Zealand catshark 170
New Zealand lanternshark 158
New Zealand whaler 176
Nicaragua shark 176
night shark 177
Nilson's deepsea dogfish 156
North Pacific shark 159
northern whaler 177
northern wobbegong 164
Notogaleus rhinophanes see
   Galeorhinus galeus 173
Notorynchus cepedianus 156
Notorynchus maculatus see *N.*
   cepedianus 156
Notorynchus pectorosus see *N.*
   cepedianus 156
nurse shark 165
nurse sharks 165
nursehound 172
oceanic whitetip shark 177
ocellated angel shark 162
odontaspididae 165
Odontaspis arenarius see *Eugomphodus*
   taurus 165
Odontaspis ferox 165
Odontaspis herbsti see *O. ferox* 165
Odontaspis noronhai 165
Odontaspis platensis see *Eugomphodus*
   taurus 165
Odontaspis taurus see *Eugomphodus*
   taurus 165
onefin catshark 171
orectolobidae 163-4
Orectolobiformes 163-5
Orectolobus japonicus 163
Orectolobus maculatus 163
Orectolobus ornatus 163
Orectolobus ornatus halei see *O. ornatus*
   163
Orectolobus tentaculatus see *Sutorectus*
   tentaculatus 164
Orectolobus wardi 164
ornate angel shark 162
ornate dogfish 157
ornate wobbegong 163
Owston's spiny dogfish 157
oxynotidae 160-1
Oxynotus bruniensis 160
Oxynotus caribbaeus 161
Oxynotus centrina 161
Oxynotus paradoxus 161
oystercrusher 162
Pacific angel shark 161

Pacific black dogfish 157
Pacific porbeagle 167
Pacific sharpnose shark 178
Pacific sleeper shark 159
pale catshark 168
Panama ghost catshark 168
Papuan epaulette shark 164
Paragaleus pectoralis 175
Paragaleus tengi 175
Parapristurus spongiceps see *Apristurus*
   spongiceps 168
parascylliidae 163
Parascyllium collare 163
Parascyllium ferrugineum 163
Parascyllium multimaculatum 163
Parascyllium variolatum 163
Parmaturus campechiensis 170
Parmaturus manis see *Apristurus*
   manis 167
Parmaturus melanobranchius 170
Parmaturus pilosus 171
Parmaturus stenseni see *Apristurus*
   stenseni 168
Parmaturus xaniurus 171
pelagic thresher 166
Pentanchus profundicolus 171
Pentanchus spongiceps see *Apristurus*
   spongiceps 168
peppered catshark 169
perlon shark 156
Physodon muelleri see *Scoliodon*
   laticaudus 179
pico bianco 178
pigeye shark 176
pigfish 162
piked dogfish 160
Platypodon gangeticus see *Glyphis*
   gangeticus 178
Pliotrema warreni 161
Plunket shark 157
polkadot catshark 171
Pondicherry shark 176
porbeagle 167
porbeagles 166-7
Poroderma africanum 171
Poroderma marleyi 171
Poroderma pantherinum 171
Port Jackson shark 162
Portuguese dogfish 157
Portuguese shark 157
prickle shark 156
prickly dogfish 160
prickly shark 156
Prionace glauca 178
pristiophoridae 161
Pristiophoriformes 161
Pristiophorus cirratus 161
Pristiophorus japonicus 161
Pristiophorus nudipinnis 161
Pristiophorus schroederi 161
proscylliidae 172
Proscyllium habereri 172
Protozygaena taylori see
   Rhizoprionodon taylori 179
Pseudocarcharias kamoharai 166
pseudocarchariidae 166
Pseudotriakidae 172
Pseudotriakis acrages see *P. microdon*
   172

QRS

Pseudotriakis acrales see *P. microdon* 172

Pseudotriakis microdon 172

Pterolamiops longimanus see *Carcharhinus longimanus* 177

puffadder shyshark 170

pygmy ribbontail catshark 172

pygmy sharks 159, 160

quagga catshark 170

ragged-tooth shark 165

redspotted catshark 171

requiem sharks 175-99

reticulated swellshark 168

Rhincodon typus see *Rhiniodon typus* 165

Rhineodon typus see *Rhiniodon typus* 165

rhiniodontidae 165

Rhiniodon typus 165

Rhizoprionodon acutus 178

Rhizoprionodon lalandii 178

Rhizoprionodon longurio 178

Rhizoprionodon oligolinx 178

Rhizoprionodon porosus 178

Rhizoprionodon taylori 179

Rhizoprionodon terraenovae 179

rig 174

river shark 176

rough hound 171

rough longnose dogfish 157

roughsharks 160-1

roughskin dogfish 157

roughskin spiny dogfish 160

roughskin spurdog 160

roughtail catshark 169

rusty carpetshark 163

saddle carpetshark 163

sailback houndshark 173

sailfin roughshark 161

sailfish shark 166

salamander shark 171

Saldanha catshark 168

salmon shark 167

sand devils 161-2

sand shark 165

sandtiger sharks 165

sandbar shark 177

sandtiger shark 165

sandy dogfish 171

sawsharks 161

sawback angel shark 161

scalloped bonnethead 179

scalloped hammerhead 179

Scapanorhynchus owstoni see *Mitsukurina owstoni* 165

school shark 173

Schroederichthys bivius 171

Schroederichthys chilensis 171

Schroederichthys maculatus 171

Schroederichthys tenuis 171

Scoliodon lalandei see *Rhizoprionodon lalandii* 178

Scoliodon laticaudus 179

Scoliodon longurio see *Rhizoprionodon longurio* 178

Scoliodon palasorra see *S. laticaudus* 179

Scoliodon palasorrah see *Rhizoprionodon oligolinx* 178

Scoliodon sorrakowa see *Rhizoprionodon acutus* 178

Scoliodon terraenovae see *Rhizoprionodon terraenovae* 179

Scoliodon walbeehmi see *Rhizonprionodon acutus* 178

scoophead 179

scoophead sharks 179

scyliorhinidae 167-72

Scyliorhinus besnardi 171

Scyliorhinus boa 171

Scyliorhinus canicula 171

Scyliorhinus capensis 171

Scyliorhinus cervigoni 171

Scyliorhinus garmani 171

Scyliorhinus haeckeli 172

Scyliorhinus hesperius 172

Scyliorhinus meadi 172

Scyliorhinus retifer 172

Scyliorhinus retifer boa see *S. boa* 171 or *S. hesperius* 172

Scyliorhinus retifer haeckelii see *S. haeckeli* 172

Scyliorhinus retifer meadi see *S. meadi* 172

Scyliorhinus retifer retifer see *S. retifer* 172

Scyliorhinus stellaris 172

Scyliorhinus torazame 172

Scyliorhinus torrei 172

Scylliogaleus quecketti 174

Scymnodalatias sherwoodi 159

Scymnodon obscurus 159

Scymnodon plunketi see *Centroscymnus plunketi* 157

Scymnodon ringens 159

Scymnodon squamulosus 159

Scymnorhinus licha see *Dalatias licha* 157

seal shark 157

sevengill shark 156

sevengill sharks 156

sharp-back shark 161

sharpfin houndshark 174

sharpnose sevengill shark 156

sharp-nosed shark 178

sharptooth houndshark 175

sharptooth smooth-hound 173

Sherwood dogfish 159

short-tail nurse shark 165

shortfin mako 166

shortnose sawshark 161

shortnose spurdog 160

shortnose velvet dogfish 157

shortspine spurdog 160

shorttail lanternshark 158

shovelnose 176

shovel-nose shark 165

shovel-nosed shark 157

shy eye shark 170

sickle shark 176

sicklefin lemon shark 178

sicklefin smooth-hound 174

sicklefin weasel shark 175

silky shark 176

silvertip shark 175

silvertip whaler 175

sixgill sawshark 161

sixgill sharks 156

skittledog 160

sleeper sharks 159-60

slender bamboo shark 164

slender catshark 171

slender smooth-hound 172

slime shark 159

slipway grey shark 176

sliteye shark 178

small-spotted catshark 171

smallbelly catshark 167

smalleye catshark 167

smalleye hammerhead 179

smalleye smooth-hound 174

smallfin catshark 168

smallfin gulper shark 156

smallmouth velvet dogfish 159

smalltail shark 177

smalltooth sand tiger 165

smalltooth thresher 166

smooth dog shark 173

smooth dogfish 173

smoothfanged shark 176

smooth hammerhead 179

smooth lanternshark 158

smooth-hound 172, 174

smooth-hounds 173-5

smoothback angel shark 162

snaggletooth shark 175

Somniosus antarcticus see *S. microcephalus* 159

Somniosus microcephalus 159

Somniosus pacificus 159

Somniosus rostratus 160

South China catshark 168

southern dogfish 156

southern lanternshark 158

southern sawshark 161

spadenose shark 179

spatulasnout catshark 168

speartooth shark 178

speckled carpetshark 164

speckled catshark 164, 170

speckled smooth-hound 174

Sphyrna blochii see *Eusphyra blochii* 179

Sphyrna corona 179

Sphyrna couardi 179

Sphyrna lewini 179

Sphyrna media 179

Sphyrna mokarran 179

Sphyrna tiburo 179

Sphyrna tudes 179

Sphyrna zygaena 179

sphyrnidae 179

Spiky Jack 160

spined pgymy shark 160

spinner shark 176

spiny dogfish 160

spiny shark 156

spitting shark 165

spongehead catshark 168

spot-tail shark 177

spotless catshark 170

spotless smooth-hound 173

spotted estuary smooth-hound 174

spotted houndshark 175

spotted ragged-tooth shark 165

spotted wobbegong 163

spurdogs 160

squalidae 156-60
Squaliformes 156-61
Squaliolus laticaudus 160
Squaliolus sarmenti see *S. laticaudus* 160
Squalus acanthias 160
Squalus americanus see *Dalatias licha* 157 or *Eugomphodus taurus* 165
Squalus asper 160
Squalus blainvillei 160
Squalus cubensis 160
Squalus japonicus 160
Squalus kirki see *S. acanthias* 160
Squalus megalops 160
Squalus melanurus 160
Squalus mitsukurii 160
Squalus rancureli 160
square-nose shark 176
Squatina aculeata 161
Squatina africana 161
Squatina argentina 161
Squatina armata see *S. californica* 161
Squatina australis 161
Squatina californica 161
Squatina dumeril 161
Squatina formosa 161
Squatina japonica 162
Squatina nebulosa 162
Squatina oculata 162
Squatina squatina 162
Squatina tergocellata 162
Squatina tergocellatoides 162
squatinidae 161-2
Squatiniformes 161-2
starry smooth-hound 173
starspotted smooth-hound 174
Stegostoma fasciatum 164
stegostomatidae 164
Stegostoma tigrinum see *S. fasciatum* 164
stellate smooth-hound 173

T

straight-tooth weasel shark 175
striped catshark 171
striped smooth-hound 173
Sutorectus tentaculatus 164
Sweet William 173, 175
swellshark 169
swellsharks 168-9
taillight shark 159
Taiwan angel shark 161
Taiwan gulper shark 156
Taiwan saddle carpetshark 163
Tasmanian carpetshark 163
Tasmanian catshark 163
Tasmanian spotted catshark 163
tasselled wobbegong 163
tawny nurse shark 165
tessellated deepwater shark 157
thintail thresher 166
Thompson's shark 157
thorndog 160
thorny lanternshark 158
thresher shark 166
thresher sharks 166
tiger catshark 170
tiger shark 177
Tommy 166
tope shark 173
topes 173-5
topesharks 173
Triaenodon obesus 179
triakidae 173-5
Triakis acutipinna 174
Triakis henlei see *Mustelus henlei* 174
Triakis maculata 175
Triakis megalopterus 175
Triakis natalensis see *T. megalopoterus* 175
Triakis scyllia see *T. scyllium* 175
Triakis scyllium 175
Triakis semifasciata 175
Triakis venusta see *Proscyllium*

U-Z

habereri 172
uptail 166
Van Rooyen's shark 176
varied catshark 163
velvet belly 158
velvet dogfish 159
vitamin shark 173
weasel shark 175
weasel sharks 175
West African catshark 171
whale shark 165
whalers 177
whiptail shark 166
whiskery shark 173
whiskery sharks 173-5
white death 166
white pointer 166
white sharks 166-7
whitecheek shark 176
white-cheeked whaler shark 176
whitefin dogfish 157
whitefin hammerhead 179
whitefin topeshark 173
whitefinned swellshark 169
whitenose shark 178
whitesaddled catshark 172
whitespotted bambooshark 164
whitespotted bullhead shark 162
whitespotted dogfish 160
whitespotted gummy shark 174
whitespotted shark 162
whitespotted smooth-hound 174
whitetip reef shark 179
whitetip sharks 177, 179
winghead shark 179
wobbegongs 163-4
yellowspotted catshark 171
Zambezi shark 176
zebra bullhead shark 162
zebra shark 164

ACKNOWLEDGEMENTS

Many people and organisations assisted in the preparation of this book. The publishers would like to thank them all, particularly:

Australian Museum, Sydney; Mr Foster Bam; Bank of England; Guy Bevan; Henri Bource; John Casey, National Marine Fisheries Service, Narragansett; Coledale Surf Club; Dr Leonard J.V. Compagno; Robert Curry; Stephanie Davenport; Wade Doak; Sara Dodds; Rod Grainger of *Bounty Hunter*; Howard Hall; Peter Huck; Robyn Hudson; Stuart Inder; Eric Love; Mitchell Library, Sydney; Museum of Applied Arts and Sciences, Sydney; National Library of Jamaica; Newcastle and Port Stephens GFC; Ocean Leather Corporation; Qantas Airways Ltd; Sean Semler; Linda Spain; State Library of New South Wales; Dr J. D. Stevens; Dr Leighton R. Taylor, Waikiki Aquarium; Dr Tony Underwood; Robert J. Woodward.

The following organisations and individuals provided photographs and gave permission for them to be reproduced.

6-7: Ron & Valerie Taylor. 8-9: Ron & Valerie Taylor. 12: Photri. 14: tl, Kathie Atkinson; others, Ron & Valerie Taylor. 16 and 17: tr, tl, Dianne R. Hughes. 20: David Doubilet. 21: United States Navy. 24: t, Bill Wood; b, Ron & Valerie Taylor. 26: Ron & Valerie Taylor. 27: United States Navy. 28: tl, Nick Otway; bl, br, Ron & Valerie Taylor. 29: Foster Bam. 30, 31: David Doubilet. 33: Ron & Valerie Taylor. 35: Ron & Valerie Taylor. 39: t, Susan Turner; b, Noel Kemp. 41: Noel Kemp. 42: tl, Noel Kemp; tr, Susan Turner. 43: t, Neg. No. 32603. Courtesy American Museum of Natural History; b, Noel Kemp. 44: S. Church. 46: t, United States Navy; b, H. Wes Pratt. 47: Stephanie Davenport. 48: National Marine Fisheries Service – Narragansett Laboratory. 49: T. Halavik. 50-51: Ron & Valerie Taylor. 52-53: Ron & Valerie Taylor. 55: t, Ron & Valerie Taylor; b, Neg. No. 29665, courtesy Dept. Library Services, American Museum of Natural History. 56: United States Navy. 57: Howard Hall. 60-61: Mansell Collection. 62: National Museum, Naples/Agenzia Fotografica Luisa Ricciarni, Milan. 63: Kunsthistorischen Museum, Vienna. 64: t, b, Nick Otway. 65: t, Mary Evans Picture Library. 67: box, all Peter Goadby. 68: Peter Goadby. 69: t, Mary Evans Picture Library; b, BBC Hulton Picture Library. 71: tl, Stuart Inder. 72, 73: Ron & Valerie Taylor. 74-75: The Photo Library. 76, 77: Ron & Valerie Taylor. 78: t, c, Ron & Valerie Taylor; b, Bill Wood. 79: Ron & Valerie Taylor. 80: Ron & Valerie Taylor. 82, 83: Ron & Valerie Taylor, 84: Ron & Valerie Taylor. 85: tl, Ron & Valerie Taylor; tr, West Australian Newspapers Ltd.; b, Mansell Collection. 86: Ron & Valerie Taylor. 88: K. J. Waddingham. 90: Henri Bource. 91: Ron & Valerie Taylor. 92: Mary Evans Picture Library. 93: Ron & Valerie Taylor. 94: Mitchell Library, Sydney. 95: Ron & Valerie Taylor. 96: Photo Index. 96-97: Bob Mossel. 98: John Fairfax & Sons Ltd. 99: Mary Evans Picture Library. 100: Metropolitan Museum of Art, Wolfe Fund, 1906. Catharine Lorillard Wolfe Collection. (06.1234). 101: t, b, Mary Evans Picture Library. 102: t, Mary Evans Picture Library; b, Ron & Valerie Taylor. 103: Mary Evans Picture Library. 104-105: Mitchell Library, Sydney. 106, 107: Ron & Valerie Taylor. 108-109: Australian Museum/Camera Lucida. 110, 111: Wade Doak. 113: National Library of Jamaica. 115: John Fairfax & Sons Ltd. 116: John Fairfax & Sons Ltd. 117: tr, Ron & Valerie Taylor. 118: John Fairfax & Sons Ltd. 119: tr, John Fairfax & Sons Ltd. 121: Brown Brothers. 122, 123: Brown Brothers. 124, 125: Brown Brothers. 126: Ron & Valerie Taylor. 128, 129: Ron & Valerie Taylor. 130, 131: Ron & Valerie Taylor. 132: b, Ron & Valerie Taylor. 132-133: Museum of Fine Arts, Boston. 134: Museum and Historical Research Section, Bank of England. 136: Ron & Valerie Taylor. 137: tr, National Marine Fisheries Service – Narragansett Laboratory. 138: Ron & Valerie Taylor. 139: t, Foster Bam; b, United States Navy. 140: United States Navy. 141: tl, United States Navy; tr, Qantas; b, Ron & Valerie Taylor. 142, 143: Ron & Valerie Taylor. 144: t, Ocean Leather Corporation; bl, br, Museum of Applied Arts and Sciences, Sydney. 145: BBC Hulton Picture Library. 146: t, c, Ron & Valerie Taylor. 148: Eric Love. 149: t, Peter Goadby; b, Ron & Valerie Taylor. 150: t, Mark Deeney. 151: t, Mark Deeney; b, David Burford, Fish Marketing Authority of NSW. 180: t, Ron & Valerie Taylor; b, New Zealand National Publicity. 181: Australian Information Service. 182: David Rodgers, NSW Department of Agriculture, Fisheries Research Institute. 183: t, United States Navy; b, Ron & Valerie Taylor. 185: Peter Goadby. Position of photographs on the page: t – top, c – centre, b – bottom, l – left, r – right.

The publishers acknowledge their indebtedness to the following books and papers that were used for reference, and in the preparation of text and illustrations.

General reference *About Sharks and Shark Attack* David H. Davies (Routledge, London, 1965); *Animal Facts and Feats* Gerald L. Wood (Guinness Superlatives, Enfield, 1982); *Blue Meridian* Peter Matthiessen (Random House, New York, 1971); *The Book of Sharks* Richard Ellis (Grosset & Dunlap, New York, 1976); *Danger Shark* Jean Campbell Butler (Robert Hale, London); *Discover* magazine, various editions (New York, 1985/86); *Fangs of the Sea* Norman Caldwell and Norman Ellison (Angus and Robertson, Sydney, 1937); *Fishes of New Guinea* Ian S. R. Monro (Dept of Agriculture, Port Moresby, 1967); *Fishes of Tasmania* P. R. Last (Tasmanian Fisheries Development Authority, Hobart, 1983); *The Fresh and Salt Water Fishes of the World* Edward C. Migdalski and George S. Fichter (Knopf, New York, 1976); *Great Shark Stories* Ron and Valerie Taylor with Peter Goadby (Collins, London, 1978); *Shark Attack* H. David Baldridge (Berkley, New York, 1974); *Shark Attack* V. M. Coppleson (Angus and Robertson, Sydney, 1958); *Shark Attack: A Program of Data Reduction and Analysis* H. David Baldridge (Mote Marine Laboratory, Sarasota, 1974); *The Great Shark Suit Experiment* Ron and Valerie Taylor (Ron Taylor Film Productions, Sydney, 1981); *Harpoon at a Venture* Gavin Maxwell (Penguin, 1984); *The Histories* Herodotus (Penguin, 1968); *Life of Sharks, The* Paul Budker (Weidenfeld, London, 1971); *Man is the Prey* (Panther, London, 1971); *Maneater Man* Colin Thiele (Rigby, Adelaide, 1979); *Megamouth – A New Species...* Leighton R. Taylor, L. J. V. Compagno and Paul J. Struhsaker (California Academy of Sciences, 1983); *Myth and Maneater* David Webster (Angus and Robertson, Sydney, 1972); *National Geographic* magazine, various editions; *Reader's Digest Book of the Great Barrier Reef* (Reader's Digest, Sydney, 1984); *Shadows in the Sea* Harold W. McCormick, Tom Allen and Captain William Young (Weathervane, New York, 1963); *The Shark Arm Case* Vince Kelly (Nelson, Sydney); *Shark Attack in Southern African Waters* Tim Wallett (Struik, Cape Town, 1983); *Shark – Killer of the Sea* Thomas Helm (Robert Hale, London, 1962); *Shark-O* P. F. O'Connor (Secker and Warburg, London, 1953); *Shark, The: Splendid Savage of the Sea* J-Y. and P. Cousteau (Cassell, London, 1970); *Shark – The Killer of the Deep* Zane Grey (ed. L. Grey) (Belmont, New York); *Sharks* (Wildlife Education, San Diego, 1983); *Sharks: An Introduction for the Amateur Naturalist* Sanford A. Moss (Prentice-Hall, New Jersey, 1984); *Sharks and Shark Deterrents* P. W. and C. Gilbert (Underwater Journal, 1973); *Sharks and Shipwrecks* Hugh Edwards (Lansdowne, Melbourne, 1975); *Sharks and Survival* Perry W. Gilbert; *Sharks – Attacks, Habits, Species* Peter Goadby (Ure Smith, Sydney, 1959); *Sharks – Killers of the Deep* Michael Bright (Omega, London, 1984); *Sharks of Australia* G. P. Whitley (Pollard, Sydney, 1981); *Sharks of the World, vols 1 and 2* Leonard J. V. Compagno (FAO, Rome, 1984); *Sharks of the World* Rodney Steel (Blandford, Poole, 1985); *Sharks – The Search for a Repellent* Theo W. Brown (Angus and Robertson, Sydney, 1973); *Trawled Fishes of Southern Indonesia and Northwestern Australia* Thomas Gloerfelt-Tarp and Patricia J. Kailola (ADAB); *The Water World* J. W. Van Dervoort (Union, New York, 1884). **The Mysterious Lives and Habits of Sharks** Brain to body-weight diagram, p 21, adapted from Moss, Sanford A., 1984. Sharks – an introduction for the amateur naturalist, Prentice-Hall, New Jersey, p 126. **The Shark's Remarkable Senses** The original work presented here was carried out at Cornell University, Ithaca, New York; the Mount Desert Island Biological Laboratory, Salsbury Cove, Maine; the Marine Biological Laboratory, Woods Hole, Massachusetts; the Lerner Marine Laboratory, Bimini, Bahamas; the Mote Marine Laboratory, Sarasota, Florida. Dr Perry W. Gilbert, the author, deeply appreciates the use of the facilities at each of these institutions. Portions of this study were supported by the US Office of Naval Research, the Mote Scientific Foundation, and the Elizabeth Mote Rose Research Fund. Critical editorial assistance of the author's wife, Claire K. Gilbert, and typing of the manuscript by Linda M. Franklin are gratefully acknowledged. References cited in the text and used in preparing illustrations: Budker, P., 1959. Les organes sensoriels cutanes des selaciens. In: Traite de zoologie, Tome XIII, Fasc. 2, p 1047. Corwin, J. T., 1978. The relation of inner ear structure to the feeding behaviour in sharks and rays. In: Scanning electron microscopy. Johari, O. (ed.), pp 1105-1112, Chicago: SEM, Inc. Daniel, J. F., 1934. The elasmobranch fishes. p 259. Univ. Cal. Press, Berkeley, California. Dijkgraff, S. and Kalmijn, A. J., 1963. Untersuchungen uber de Funktion der Lorenzinischen Ampullen an Haifishchen. Z. Vergl. Physiol., vol. 47, pp 438-456. Gilbert, P. W., 1962. The behaviour of sharks. Scient. Amer., vol. 207, pp 60-68. Gilbert, P. W. Biology and behaviour of sharks. Endeavour, New Series, vol. 8, no.4, pp 179-187. Gilbert, P. W., Hodgson, E. S., and Mathewson, R. F., 1964. Electroencephalograms of sharks. Science, vol. 145, pp 949-951. Hodgson, E. S. and Mathewson, R. F. (eds.), 1978. Sensory biology of sharks, skates and rays. pp i – ix, 1-666, US Dept of the Navy, Office of Naval Research. Kalmijn, A. J., 1971. The electric sense of sharks and rays. Journ. Exper. Biol., vol. 55, pp 371-383. Kalmijn, A. J., 1974. The detection of electric fields from inanimate and animate sources other than electric organs. In: Handbook of sensory physiology. Fossard, A. (ed.), vol. III, pp 147-200, Berlin, Heidelberg, New York: Springer-Verlag. Murray, R. W., 1960. Electrical sensitivity of the ampullae of Lorenzini. Nature, vol. 187, p 957. Murray, R. W., 1962. The response of the ampullae of Lorenzini of elasmobranchs to electrical stimulation. Journ. Exper. Biol., vol. 39, pp 119-128. Popper, A. N. and Fay, R. R., 1977. Structure and function of the elasmobranch auditory system. Amer. Zool., vol. 17, pp 443-452. Tester, A. L., Kendall, J. I., and Milisen, W. B., 1972. Morphology of the ear of the shark genus *Carcharhinus* with particular reference to the macula neglecta. Pacific Science, vol. 26, pp 264-274. **Inhabitants of Ancient Seas** Bauchot, R., Platel, R. & Ridet, J-M., 1976. Brain-body weight relationships in Selachii. Copeia, 1976, no. 2, pp 305-310. Duffin, C. J. & Ward, D. J., 1983. Neoselachian sharks' teeth from the lower Carboniferous of Britain and the lower Permian of the USA. Palaeontology, 26, pp 93-110. Karatajute-Talimaa, V. N., 1973. *Elegestolepis grossi* gen. et sp. nov., ein neuer Typ Placoidschuppe aus dem oberen Silur der Tuwa. Palaeontographica, A, 143, 35-50. Maisey, J. G., 1982. The anatomy and interrelationships of Mesozoic Hybodont sharks. American Museum Novitates 2724, 1-48. Maisey, J. G., 1984. Higher elasmobranch phylogeny and biostratigraphy. Zoological Journal of the Linnean Society 82, pp 33-54. Maisey, J. G. & Wolfram, K. E., 1984. 'Notidanus'. In N. Eldridge & S. Stanley (eds), 'Living Fossils', Springer-Verlag, pp 170-180. Mader, H., 1986. Schuppen und Zahne von Acanthodiern und Elasmobranchiern aus dem Unter-Devon Spaniens (Pisces). Gottingen Arbieten zur Geologie und Palaontologie 28, pp 1-58. Thies, D. & Reif, W-E., 1985. Phylogeny and evolutionary ecology of Mesozoic Neoselachii. Neues Jahrbuch fur Geologie und Palaontologie, Abhandlungen pp 169, 333-361. Turner, S., 1982b. Middle Palaeozoic elasmobranch remains from Australia. Journal of Vertebrate Paleontology 2 (2), pp 117-131. Young G. C., 1982. Devonian sharks from south-eastern Australia and Antarctica. Palaeontology, 25, pp 817-843. Zangerl, R., 1981. Chondrichthyes I, Paleozoic Elasmobranchii. Vol. 3A Handbook of Palaeoichthyology (ed. H-P. Schultze), Gustav Fischer-Verlag, Stuttgart/New York, pp 115. **Sharks in the wild** Information on shark movements in the Atlantic Ocean from Casey, J. et al. The shark tagger, 1977-1985. National Marine Fisheries Service, Narragansett, R. I. Stevens, J. D. 1976. First results of shark tagging in the north-east Atlantic, 1972-1975. J. Mar. Biol. Ass. UK 56, pp 929-937. Information on Australian tagging results from Pepperell, J. Gamefish tagging 1981-82 – 1984-85. New South Wales Dept. Agric., Fish. Div. Tracking of white shark using sonic tag from: Compagno, L. J. V., 1984. FAO species catalogue. Vol. 4 Sharks of the World. FIR/S125 Vol. 4. Part 1. Segregation of sexes from: Bass, A. J., D'Aubrey, J. D. and Kistnasamy, N. 1976. Sharks of the east coast of southern Africa. SA Ass. Mar, Biol. Res. ORI Invest. Rep. No. 45. Growth and ageing of sharks: Prince, E. D. and Pulos, L. M. (eds), 1983. Proceedings of the international workshop on age determination of oceanic pelagic fishes. NOAA Tech. Rep. NMFS 8. Growth rate of sandbar sharks – Casey et al. (above). Growth rate of Galapagos whalers: Bass, A. J. and Smith, J. L. B. 1977. Long-term recoveries of tagged sharks. Copeia, No. 3. **Sharks of the World** Diagram on p 155 adapted from Compagno, Leonard J. V., 1984 Sharks of the World. p 3, FAO, Rome.

Typesetting by Smithys Ltd, NSW, Australia
Colour reproduction by Colourscan Co. Pte Ltd, Singapore
Paper supplied by Kymme Star Paper Co. (F/E), Hong Kong

Printed and bound in 1986 by
Dai Nippon Printing Co. (HK) Ltd, Hong Kong for
Reader's Digest Services Pty Limited (Inc in NSW),
26-32 Waterloo Street, Surry Hills,
NSW 2010, Australia.